T0224987

SpringerBriefs in Computer Science

SpringerBriefs present concise summaries of cutting-edge research and practical applications across a wide spectrum of fields. Featuring compact volumes of 50 to 125 pages, the series covers a range of content from professional to academic.

Typical topics might include:

- A timely report of state-of-the art analytical techniques
- A bridge between new research results, as published in journal articles, and a contextual literature review
- A snapshot of a hot or emerging topic
- An in-depth case study or clinical example
- A presentation of core concepts that students must understand in order to make independent contributions

Briefs allow authors to present their ideas and readers to absorb them with minimal time investment. Briefs will be published as part of Springer's eBook collection, with millions of users worldwide. In addition, Briefs will be available for individual print and electronic purchase. Briefs are characterized by fast, global electronic dissemination, standard publishing contracts, easy-to-use manuscript preparation and formatting guidelines, and expedited production schedules. We aim for publication 8–12 weeks after acceptance. Both solicited and unsolicited manuscripts are considered for publication in this series.

**Indexing: This series is indexed in Scopus, Ei-Compendex, and zbMATH **

Bernard Chen

Wineinformatics

A New Data Science Application

 Springer

Bernard Chen
University of Central Arkansas
Conway, AR, USA

ISSN 2191-5768   ISSN 2191-5776   (electronic)
SpringerBriefs in Computer Science
ISBN 978-981-19-7368-0   ISBN 978-981-19-7369-7   (eBook)
https://doi.org/10.1007/978-981-19-7369-7

This Springer imprint is published by the registered company Springer Nature Singapore Pte Ltd.
The registered company address is: 152 Beach Road, #21-01/04 Gateway East, Singapore 189721,
Singapore

Preface

Wine is an ancient beverage produced and enjoyed by humans for several thousand years, and it has become more popular and affordable nowadays. Hundred million hectoliters of wine was produced across more than 30 countries each year. During the grape growing, wine making, and wine evaluating process, lots of data are generated and stored. With the right data mining/data science techniques, hidden knowledge and information can be retrieved from the large amount of wine-related data.

This book introduces a new data science application domain named Wineinformatics which combines wine-related data with data mining/data science techniques to discover new types of knowledge in wine. Supervised learning, which is one of four learning types of algorithms, applied on professional wine reviews is the main focus used in this book. With the development and utilization of the Computational Wine Wheel, large volumes of wine reviews in human language format are converted into computer understandable binary encodings. New information can be mined through a properly designed dataset to answer some tough and interesting questions on wine.

Chapter 1 provides an introduction to data science and Wineinformatics. Chapter 2 introduces the development and usage of the Computational Wine Wheel and the three datasets used in this book. Chapter 3 presents a set of experiments to answer the question of *"How does a wine achieve 90+ scores?"* using classification algorithms. Chapter 4 discusses how to evaluate wine judges and figure out the question of *"Are wine reviewers reliable and consistent?"* Chapter 5 targets the question of *"Can actual wine grade and price be predicted through their reviews?"* using regression analysis. Chapter 6 introduces newer and more complicated computer science techniques, Multi-Label and Multi-Target, to Wineinformatics. These techniques are useful to work on the problem of *"Can wine grade, price and region be predicted altogether with higher accuracy?"*

Chapter 7 discusses how to extract more information from the Computational Wine Wheel to respond to the question of *"How can computers understand wine reviews even more?"* Finally, Chapter 8 draws the conclusion and provides some of many intriguing future works.

Conway, AR, USA   Bernard Chen

Contents

Chapter 1
Introduction

Abstract This chapter provides an introduction to data science and Wineinformatics.

1.1 Data Science

Science has its origins in attempts to learn the world in an empirically verified manner. To understand the world, one relies on testing it through **data**. Each data point is a discrete record or a property of events that occurred. With the rise of the Internet, data has become abundant in various topics associate with different formats. At the end of 2020, it was estimated that approximately 1.7 MB of new information was generated per second for all humans on Earth. This trend has made it extremely important to know how to extract useful meaning from large amount of data. **Data science** is a successful study that includes different techniques and theories from different fields, including mathematics, computer science, economics, and business administration to gain unique insight from data related to its domain.

Data science can be applied in any filed; however, to extract and discover novel knowledge, the study requires a domain knowledge such as finance, healthcare, corporate services, media and communications, software and IT services. Within this emerging field, there are several different types of learning algorithms that provide utility. Specifically, there are four major types of learning algorithms based on the problem, the input/output, and methods: supervised learning [1], unsupervised learning [2], semi-supervised learning [3], and reinforced learning [4]. Generally speaking, supervised learning builds up models from known data for predicting unknown dataset; clustering groups similar items from their attributes to discovery patterns; semi-supervised learning combines both supervised and unsupervised (clustering) learning methods; and reinforced learning mimics human learning activates to maximize rewards. All of these methods are useful for discovering interesting information from large amounts of data within specific application domains.

This book uses **wine** as the application domain and demonstrates how **classification** algorithms can be applied to predict and understand wine scores and reviews as well as evaluate the wine judges.

1.2 Wineinformatics

Wine as a beverage with alcohol has been produced and developed for several thousands of years. Wine has remained popular and even more affordable in modern times. According to OIV (International Organization of Vine and Wine), who is the world's authority on wine statistics, in 2018, 293 million hectoliters of wine were produced across 36 countries; this constitutes a 17% increase in wine production from 2017 to 2018 [5]. Numerous grape types combine with different wine making processes create endless number of flavor and style wines; as not many of consumers are wine experts, wine reviews and rankings produced by the experts, such as Robert Parker [6], James Suckling [7], and wine review magazines, such as Wine Spectator [8], Wine Enthusiast [9], and Decanter [10], are important heuristics to consumers' decision-making. Wine reviews benefits not only consumers but also wine makers and distributors. Wine makers can gain valuable information and knowledge from the expert reviews by knowing what factors contribute the most to quality, as determined by rankings. Wine distributors can either determine representative wines of regions to increase the breadth of the wine list or select distinct wines in a region to focus the depth of the wine list.

Hundreds of thousands wine reviews are available in each wine expert's database. Both Robert Parker's Wine Advocator and Wine Spectator currently have more than 400,000 wine reviews available. Therefore, millions of wine reviews written by experts stored in human language format across multiple decades are available to help researchers understand wine. These reviews serve as the final verdicts to reflect wine maker's hard work, terroir, vintage, weather impacts and many other factors. How to mine and discover useful information from the large amount of available wine reviews for wine producers, distributors, and consumers is a major task to the current data science challenge.

Wineinformatics incorporates data science and wine-related datasets produced from the process of grape growing, wine making and wine evaluating to discover useful information for wine producers, distributors, and consumers [11, 12]. To describe the end product of wine, two type of data resulting from wine evaluation process: Physicochemical laboratory data and wine reviews. Physicochemical laboratory data usually relates to the physicochemical composition analysis [13], such as acidity, residual sugar, alcohol, etc., to characterize wine. Physicochemical information is more useful for the winemaking process and usually contains smaller amount of data (less than 200 wines) due to the cost of the experiments. Wine reviews are produced by sommeliers, who are people who specialize in wine. These wine reviews usually include aroma, flavors, tannins, weight, finish, appearance, and the interactions related to these wine sensations [14].

Unlike physicochemical laboratory data can be easily read and applied analytics by computers, wine reviews require the works of **natural language processing** and a degree of human bias for computers to understand the human language format reviews. Nonetheless, wine reviews have millions of potential data that are available to researchers with little cost and can provide useful information to broader audiences. Therefore, this book focuses on how to use the **Computational Wine Wheel** to accurately capture keywords, including not only flavors but also non-flavor notes, which always appear in the wine reviews [15, 16]. With the help of the Computational Wine Wheel, region specific or research topics oriented wine reviews can be collected and processed as the clean dataset for data science research for answering the questions such as "What makes wine achieve a 90+ rating and considered as an outstanding wine?", "What are the shared similarities amongst groups of wine?", "What characteristics of classic (95+) wines show in 21 century Bordeaux?", "Are wine reviewers reliable and consistent?"

1.3 Book Overview

Each Chapter of this book answers one major question related to Wineinformatics. Chapter 2 discuss how to make computers understand wine reviews through the Computational Wine Wheel. Different datasets with specific purposes were developed for the following Chapters. Chapter 3 unveils how does a wine achieve 90+ scores through classification models with different evaluation methods. Chapter 4 answers the question of are the wine judges reliable and consistent with SVM and Naïve Bayes classification model. Chapter 5 makes predictions on actual wine grade and price via regression model. Chapter 6 utilizes multi-class, multi-label and multi-target approaches to predict wine grade, price and region altogether and discover subtle relationship between predicted labels. Chapter 7 loops back to the Computational Wine Wheel and seeks the answer for How can computers understand wine reviews even more? Chapter 8 draw the conclusion and points out several important future works in Wineinformatics.

References

1. Muhammad, I., Yan, Z.: Supervised machine learning approaches: a survey. ICTACT J. Soft Comput. **5**, 946–952 (2015)
2. Khanum, M., Mahboob, T., Imtiaz, W., Ghafoor, H.A., Sehar, R.: A survey on unsupervised machine learning algorithms for automation, classification and maintenance. Int. J. Comput. Appl. **119**, 34–39 (2015)
3. Van Engelen, J.E., Hoos, H.H.: A survey on semi-supervised learning. Mach. Learn. **109**, 373–440 (2020)
4. Padakandla, S.: A survey of reinforcement learning algorithms for dynamically varying environments. ACM Comput. Surv. **54**, 1–25 (2021)

5. Karlsson, P.: World wine production reaches record level in 2018, consumption is stable. BKWine Magazine. https://www.bkwine.com/features/more/world-wine-production-reaches-record-level-2018-consumptionstable/ (2019). Accessed 1 Jan 2022

6. Parker, R.: Wine Advocate. https://www.robertparker.com/. Accessed 1 Jan 2022

7. Suckling, J.: Wine ratings. https://www.jamessuckling.com/tag/wine-ratings/. Accessed 1 Jan 2022

8. Wine Spectator. https://www.winespectator.com. Accessed 1 Jan 2022

9. Wine Enthusiast. https://www.wineenthusiast.com/. Accessed 1 Jan 2022

10. Decanter. https://www.decanter.com/. Accessed 1 Jan 2022

11. Chen, B., Velchev, V., Palmer, J., Atkison, T.: Wineinformatics: a quantitative analysis of wine reviewers. Fermentation. **4**, 82 (2018)

12. Palmer, J., Chen, B.: Wineinformatics: regression on the grade and price of wines through their sensory attributes. Fermentation. **4**, 84 (2018)

13. Cortez, P., Cerdeira, A., Almeida, F., Matos, T., Reis, J.: Modeling wine preferences by data mining from physicochemical properties. Decis. Support. Syst. **47**, 547–553 (2009)

14. Edelmann, A., Diewok, J., Schuster, K.C., Lendl, B.: Rapid method for the discrimination of red wine cultivars based on mid-infrared spectroscopy of phenolic wine extracts. J. Agric. Food Chem. **49**, 1139–1145 (2001)

15. Chen, B., Rhodes, C., Crawford, A., Hambuchen, L.: Wineinformatics: Applying data mining on wine sensory reviews processed by the computational wine wheel. In: Proceedings of the 2014 IEEE International Conference on Data Mining Workshop, Shenzhen, China, 14 December 2014, pp. 142–149

16. Chen, B., Rhodes, C., Yu, A., Velchev, V.: The computational wine wheel 2.0 and the TriMax triclustering in wineinformatics. In: Industrial Conference on Data Mining, pp. 223–238. Springer, Cham (2016)

Chapter 2
Data Collection and Preprocessing

Abstract Since wine reviews are written in human language format, computers cannot understand the meanings without proper preprocessing works. Therefore, *"How to make computers understand wine reviews?"* is the main problem to be solved in this chapter. How to develop and use the Computational Wine Wheel is carefully described. Three datasets with different focuses are developed and used throughout the book based on the introduced method.

2.1 Wine Sensory Reviews

Wine testing is a delicate process since wine are examined and evaluated not only for its tasting quality, but for physical appearance and physiochemical properties as well. Wine reviewer needs to evaluate the appearance of the wine, the smells in the glass before tasting, the different sensations once tasted, and finally how the wine finishes with its aftertaste. The verdict of the wine is based on how complex the wine is, how much potential it has for aging for drinkability, and if there are any faults present. The experience required can be expansive as any given wine needs to be carefully assessed within comparable wine standards according to its wine making process, region, varietal, and style. Several prestigious wine expert or magazines, such as Wine Spectator, Wine Enthusiast, Wine Advocate and Decanter, review wines periodically and publish their reviews on their magazine and online. In order to provide a concrete example of a professional wine review, below is the wine tasting result of Wine Spectator's 2014 wine of the year review as an example.

> *Dow's Vintage Port 2011 99pts*
> *Powerful, refined and luscious, with a surplus of dark plum, kirsch and cassis flavors that are unctuous and long. Shows plenty of grip, presenting a long, full finish, filled with Asian spice and raspberry tart accents. Rich and chocolaty. One for the ages. Best from 2030 through 2060.*

B. Chen, *Wineinformatics*, SpringerBriefs in Computer Science,
https://doi.org/10.1007/978-981-19-7369-7_2

2.2   Wine Spectator

Among various prestigious wine magazines, Wine Spectator can be considered as an easier data source to start aggregating wine reviews because of their strong on-line wine review search database and consistent wine reviews. These reviews are mostly comprised of specific tasting notes and observations while avoiding superfluous anecdotes and non-related information. Wine Spectator reviews more than 15,000 wines per year and all tastings are conducted in private, under controlled conditions. The official testing process is usually held in a blind testing setup, which means the tasters are told only the general type of wine (varietal and/or region) and the vintage; no information about the winery or the price of the wine is available to the tasters while they are tasting. For each reviewed wine, a rating within a 100-points scale is given to reflect how highly their reviewers regard each wine relative to other wines in its category and potential quality. The score summarizes a wine's overall quality, while the testing note describes the wine's style and character. The reviewer will also blind test wine multiple times to ensure the consistency of the score. The overall rating reflects the following information recommended by Wine Spectator about the wine [1]:

95–100   Classic: *a great wine*
90–94   *Outstanding: a wine of superior character and style*
85–89   *Very good: a wine with special qualities*
80–84   *Good: a solid, well-made wine*
75–79   *Mediocre: a drinkable wine that may have minor flows*
50–74   *Not recommended*

2.3   Computational Wine Wheel

The tasting notes given in a review are very important as they describe the heart and soul of a wine. Even without knowing the producer or varietal, a well-described review can adequately sway a potential consumer into a purchase. However, computers cannot read and understand wine reviews as human do. In the review example listed above, the characteristics or the attributes of the wine are neatly stated without much confusion to what constitutes a proper wine tasting note in human language format. For computer to understand the reviews, the researchers need to preprocess the review and pickup important keywords as attributes; this step is called natural language processing. For example, to manually process the review, all the terms that are bold will be extracted and considered characteristics of the wine and discard non-relevant words:

> **Dow's Vintage Port 2011**
> **Powerful, refined** and **luscious,** with a surplus of **dark plum, kirsch** and **cassis** flavors that are **unctuous** and **long.** Shows plenty of **grip,** presenting a **long, full finish,** filled with

*Asian spice and raspberry tart accents. **Rich** and **chocolaty**. One for the ages. Best from 2030 through 2060.*

The bolded words are considered as attributes, which range from actual **savory** properties, such as "chocolate" and "Asian spice", body of the wine, such as "long", to **adjective** properties, such as "powerful" and "refined." However, the data used in Wineinformatics research usually comes with thousands to hundreds of thousands, it is almost impossible manually extract attributes from all targeted reviews; therefore, one of the most important research in Wineinformatics to extract as many key attributes as possible from these professional wine reviews automatically so that the computers can understand wine reviews written in human language format. To achieve the goal, two important existing researches need to be addressed: Wine Aroma Wheel and the Bag of Words (BoW) technique.

2.3.1 Wine Aroma Wheel

The Wine Aroma Wheel, developed by Ann C. Nobel [2, 3], is a great way for wine lovers to get a look at the numerous fragrances and flavors found in most wines and is composed of 12 categories of overall wine aromas someone might experience when tasting a wine. Each category has different number of subcategories and each subcategory maps with several different flavors, scents and aromatic qualities found in most red and white wines, regardless of the grape variety. An example of this would be the FRUITY ->(TREE) FRUIT ->APPLE attribute. The full list of wine aroma wheel is available in [3]. Although Wine Aroma Wheel is good for human to study wine, it is not sufficient for computer to understand the review since the aroma wheel describes only actual savory attributes; the **adjectives** and wine **body** attributes are not included. If we map Novel's wine aroma wheel to the previous example, the processed wine savory review will be:

Dow's Vintage Port 2011
*Powerful, refined and luscious, with a surplus of **dark plum, kirsch** and **cassis** flavors that are unctuous and long. Shows plenty of grip, presenting a long, full finish, filled with **Asian spice** and **raspberry tart** accents. Rich and **chocolaty**. One for the ages. Best from 2030 through 2060.*

Compare with the review example, the ideal attribute extraction process should include the review's key attributes into the three mentioned categories: savory, body, and descriptive showed below:

Dow's Vintage Port 2011

*Powerful!, refined! and luscious!, with a surplus of **dark plum***, **kirsch*** and **cassis*** flavors that are unctuous! and long!. Shows plenty of grip+, presenting a long+, full finish+, filled with **Asian spice*** and **raspberry tart*** accents. Rich! and **chocolaty***. One for the ages. Best from 2030 through 2060.*

For this review, **red(*)** words indicate specific flavors and aromas that could possibly be found on Nobel's wine aroma wheel. Orange(+) words indicate traits corresponding to the physical wine itself like its body and finish. That is, how the wine feels physically to a taster. Lastly, blue(!) words indicate subjective adjectives used by the taste to describe the overall wine.

2.3.2   Bag of Words in Natural Language Processing

The bag of words (BoW) technique provides a feature representation of free-form text that can be used by machine learning algorithms for natural language processing (NLP) [4]. Natural language processing is a methodology designed to extract concepts and meaning from human-generated unstructured (free-form) text and convert those into numerical representation so that machine learning techniques can be applied. Bag of words technique in NLP usually consist with several stages of preprocessing including tokenization, removal of stop words, token normalization and creation of a master dictionary [4, 5]. Tokenization is a fundamental step that deals with breaking texts into phrases, sentences or words and stores them in a list. To minimize the computational complexity, BoW also attempts to discard redundant words or stop words, such as "a", "an", "the", "to", "for" etc. Token normalization normalized identical or almost the same meaning words into single token, such as "goes" and "going" are normalized into "go". Creation of a mater dictionary is the final step to memorize all efforts and store the preprocessed works into a "dictionary" for the usage in the same research domain.

Another important concept used in the BoW and NLP related to this research is the N-gram [6, 7] approach. N-grams are continuous sequences of words or symbols or tokens in a document, it is also known as the neighboring sequences of items in the document. When n = 1, also known as the unigram, only the individual words is captured in a sentence; when n = 2, also called bigrams, two words right next to each other are bagged during the tokenized process; when n = 3, also named trigrams, three continuous words are bagged during the tokenized process; when n > 3, it is usually referred to as four grams or five grams and so on. N-gram is important to all NLP researches since almost all natural language documents uses combination words, such as "dark plum", "raspberry tart" in the wine review example.

2.3.3   Development of the Computational Wine Wheel (CWW)

The pilot study of the Computational Wine Wheel [8] started with a short list of 100 wines included in Wine Spectator's "Top 100 Wines of 2011". The goal is to capture all important attributes demonstrated by representative wines of the year. After carefully tokenized all keywords and created a list of stop words, 547 distinct attributes (tokens) in the form of unigram, bigram and trigram were found. Besides actual biological flavor attributes, anything corresponding to a wine's physical structure, including things like acidity, body, structure, weight, tannins, and finish is tokenized. These are properties of wine that a taster will physically taste or feel, such as how acidic the wine tastes or how well the wine coats the tongue. Lastly, generic or subjective terminologies described by wine reviewers, such as "vivid" or "beautiful" are recorded.

The next step, token normalization merged with the concept of Wine Aroma Wheel was carried out with the hierarchy of category, subcategory, original token and normalized token. For example, FRESHLY-CUT APPLE, RIPE APPLE, and APPLE are three distinct original token normalized into APPLE normalized token, while all of them fall into FRUTY category and TREE FRUIT subcategory; while GREEN APPLE fall into the same FRUTY category and TREE FRUIT subcategory, but the flavor is distinct enough to stand out on its own so that a GREEN APPLE normalized token was created. After went through the complete normalized process, 376 normalized token were created based on 547 original token. The end product of the whole process created the very first version of the Computational Wine Wheel, which is the first master dictionary in BoW technique. Table 2.1 shows the final categories and subcategories used in the first Computational Wine Wheel. Most of the category and subcategory followed Wine Aroma Wheel except the addition of OVERALL category and corresponding ACIDITY, BODY, FINISH, FLAVOR/ DESCRIPTORS, STRUCTURE and TANNINS subcategories. For each combination, it also shows the distinct count of attributes versus normalized attributes.

As shown in Table 2.1, most specific fruits and flavors were not affected a greatly by the normalization process. It was the opinionated observations in the FLAVOR/ DESCRIPTORS subcategory that resulted in the most change. This was because we needed to make sure attributes such as DELICIOUS, DELICIOUSLY, and DELI-CIOUSNESS were all normalized into DELICIOUS instead of treating them differently.

Based on the foundation of the CWW, the Computational Wine Wheel 2.0 was developed based on Wine Spectator's Top 100 Wines from 2003 to 2013 for a much more comprehensive master dictionary generation following by the same BoW process. The Computational Wine Wheel 2.0 ended up with 14 distinct categories and a total of 34 distinct subcategories, which includes two additional CATEGORY: PUNGENT and OXIDIZED. Table 2.2 demonstrates the final categories and subcategories used in Computational Wine Wheel 2.0.

CWW 2.0 contains 1881 original token and 985 normalized token. This version of the CWW also unified the plurals problem such as "BLUEBERRY" and "BLUBERRIES" should be treated as the same original token as well as normalized token. Table 2.3 provides a detail comparison between the original CWW and CWW2.0.

Table 2.1 Categorical Summary of all Wine Attributes in CWW

Category	Subcategory	Original	Normalized
CARAMEL	CARAMEL	9	7
CHEMICAL	PETROLEUM	3	1
EARTHY	EARTHY	18	2
FLORAL	FLORAL	15	15
FRUITY	BERRY	18	15
	CITRUS	11	11
	DRIED FRUIT	21	21
	FRUIT	5	4
	OTHER	7	7
	TREE FRUIT	12	9
	TROPICAL FRUIT	15	11
HERBS/VEGETABLES	CANNED/COOKED	7	7
	DRIED	6	6
	FRESH	15	12
MEAT	MEAT	1	1
MICROBIOLOGICAL	LACTIC	3	2
	YEASTY	3	3
NUTTY	NUTTY	3	3
OVERALL	ACIDITY	14	3
	BODY	17	10
	FINISH	50	6
	FLAVOR/DESCRIPTORS	217	179
	STRUCTURE	9	2
	TANNINS	24	3
SPICY	SPICE	26	21
WOODY	BURNED	11	8
	PHENOLIC	1	1
	RESINOUS	6	6

2.3.4 How to Use the CWW 2.0

To clarify the usage of the CWW, an extremely simplified CWW example, with only 2 CATEGORY, 3 SUBCATEGORY, 6 original token and 4 normalized token, is given in Table 2.4.

Here is the process of how to apply the simplified CWW showed in Table 2.4 on the **Dow's Vintage Port 2011** wine review: The very first step is to use the words in the Specific Attribute column, which is the third column in Table 2.2, to scan the review starting with the longest number of n in n-gram token. Since the largest n in the example is 2, Raspberry Tart, Dark Plum, and Rich Aromas will be scanned on the target wine review. For every word scan, if hit, the wine will have a positive attribute in the corresponding Normalized Attribute and remove the word from the review. Therefore, after the scan of "Raspberry Tart", the program got a hit from the

Table 2.2 Categorical Summary of all Wine Attributes in CWW2.0

CATEGORY_NAME	SUBCATEGORY_NAME	SPECIFIC_NAME	NORMALIZED_NAME
CARAMEL	CARAMEL	71	40
CHEMICAL	PETROLEUM	9	5
	SULFUR	11	10
	PUNGENT	4	3
EARTHY	EARTHY	72	31
	MOLDY	2	2
FLORAL	FLORAL	61	39
FRUITY	BERRY	49	28
	CITRUS	37	23
	DRIED FRUIT	67	60
	FRUIT	22	9
	OTHER	25	18
	TREE FRUIT	39	31
	TROPICAL FRUIT	48	27
FRESH	FRESH	41	29
	DRIED	25	21
	CANNED/COOKED	16	15
MEAT	MEAT	25	13
MICROBIOLOGICAL	YEASTY	5	4
	LACTIC	14	6
NUTTY	NUTTY	25	15
OVERALL	TANNINS	90	4
	BODY	50	23
	STRUCTURE	40	2
	ACIDITY	40	3

(continued)

Table 2.2 (continued)

CATEGORY_NAME	SUBCATEGORY_NAME	SPECIFIC_NAME	NORMALIZED_NAME
	FINISH	184	5
	FLAVOR/DESCRIPTORS	649	432
OXIDIZED	OXIDIZED	1	1
PUNGENT	HOT	3	2
	COLD	1	1
SPICY	SPICE	83	44
WOODY	RESINOUS	24	9
	PHENOLIC	6	4
	BURNED	47	26

Table 2.3 Comparison of the old and new Computational Wine Wheel

	The Computational Wine Wheel 2.0	The Computational Wine Wheel
Data source	2003–2013 top 100 wines from wine spectator	2011 top 100 wines from wine spectator
Categories	14	12
Subcategories	34	28
Specific terms	1881	635
Normalized attributes	985	444
Plurals	Yes	No

Table 2.4 Simplified computational wine wheel

Category	Subcategory	Original token	Normalize token
FRUITY	BERRY	RASPBERRY	RASPBERRY
FRUITY	BERRY	RASPBERRY TART	RASPBERRY TART
FRUITY	TROPICAL FRUIT	DARK PLUM	PLUM
FRUITY	TROPICAL FRUIT	PLUM	PLUM
OVERALL	FLAVOR/DESCRIPTORS	RICH	RICH
OVERALL	FLAVOR/DESCRIPTORS	RICH AROMAS	RICH

Table 2.5 Attributes of the processed wine example

RASPBERRY	RASPBERRY TART	PLUM	RICH
0	1	1	1

wine review; the wine will have a positive value of "Raspberry Tart" attribute. After the scan of "Dark Plum", the program got a hit from the review; the wine will have another positive value of "Plum" attribute. After the scan of "Rich Aromas", which cause a miss; the wine will have a negative value of "Rich" attribute.

Once the highest n of n-gram is processed, we scan the next n of n-gram; in this example, we scan the single word specific token with the same logic. Table 2.5 represents the Dow's Vintage Port 2011's wine attributes in binary format after the process mentioned above. Please note that the RASPERRY attribute is still negative since we delete the word "RASBPERRY TART" from the review during the first scan. The readers may notice that many important attributes in the example are NOT included, such as ASIAN SPICE, CHOCHLATY... etc. It is because the computational wine wheel is the simplified version. The more SPECIFIC and NORMALIZED attributes included in the computational wine wheel, the more attributes can be picked up from the wine reviews to produce more accurate results. This is also the main reason that the Computational Wine Wheel 2.0 was studied and proposed to provide higher quality of natural language processing on wine reviews.

2.4 Dataset

The quality of the wine is based on various influences; however, two of the most well know and probably most important factors are soil and weather. Soil (or terror) reflects the characteristics of the region and depend on the composition of the soil. Weather controls the quality of the grape production and it changes every year. To recognize and study both factors in Wineinformatics, different dataset with various region and vintage specific wines can be collected and processed by the Computational Wine Wheel for machine learning analysis. In this book, three datasets were used.

2.4.1 Dataset 1: The Big Dataset

The first dataset is a collection of generic wines: all wines all over the world with vintage from 2006 to 2015 that received 80+ scores in Wine Spectator were collected for this dataset. The Big dataset ends up with over 107,000 wines in total. Wine Spectator has 10 reviewers and each one has their focus of wine regions, more detail information about their reviewers will be discussed in Chap. 4.

Figure 2.1 visualizes the most common keywords found in the data set processed by the Computational Wine Wheel. The figure demonstrates a huge variety of words describing wine in flavor.

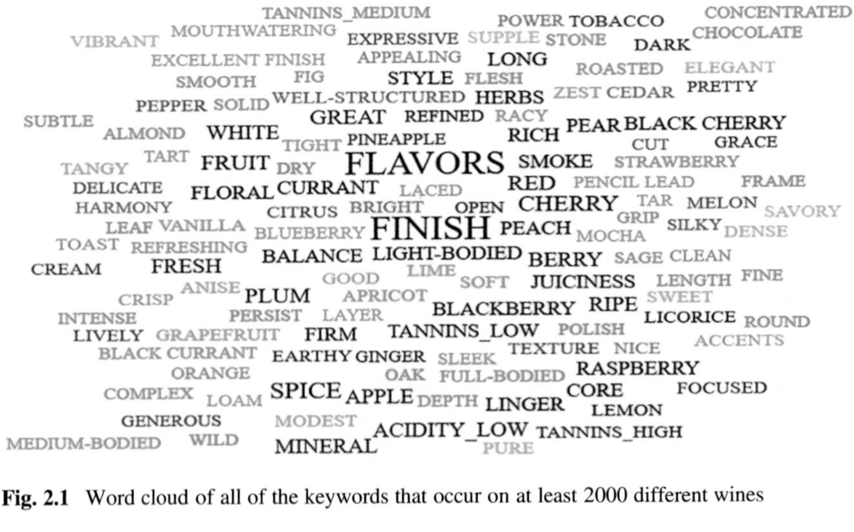

Fig. 2.1 Word cloud of all of the keywords that occur on at least 2000 different wines

2.4.2   Dataset 2: Twenty-First Century Bordeaux Dataset

The second dataset collects all Bordeaux wines made in the twenty-first century from year 2000 to 2016. A total of 14,349 wines has been collected. There are 4263 90+ wines and 10,086 89− wines. The number of 89− wines is more than 90+ wines. The score distribution is given in Fig. 2.2a. Most wines score between 86 and 90. Therefore, they fall into the category of "Very Good" wine. In Fig. 2.2b, the line chart is used to represent the trend of number of wines has been reviewed in each year. The chart also reflects the quality of vintages. More than 1200 wines were reviewed in 2009 and 2010, which indicates that 2009 and 2010 are good vintages in Bordeaux. Wine makers are more willing to send their wines to be reviewed if their wines are good.

2.4.3   Dataset 3: Twenty-First Century Elite Bordeaux

The third dataset collected all available wine reviews listed in Bordeaux Wine Official Classification, made in the twenty-first century from year 2000 to 2016. Therefore, Dataset 3 is a subset of dataset 2. A total of 1359 wines has been collected. This dataset contains 882 90+ wines and 477 89− wines. The score distribution is given in Fig. 2.3a. Unlike the data distribution of the second dataset, which has much more 89− wines than 90+ wines, in Wine Spectator, the wines selected in this research are elite choices based on Bordeaux Wine Official Classification in 1855. Therefore, classic (95+ points) and outstanding (90–94 points) wines are the majority of this dataset. The number of wines has been reviewed annually is given in Fig. 2.3b. Since Bordeaux Wine Official Classification in 1855 is a famous collection of Bordeaux wines, wine makers send their wine for review almost every year. Therefore, the line chart remains stable, which is very different from Fig. 2.2b. Regardless, some wines listed in Bordeaux Wine Official

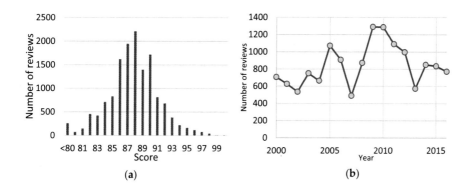

Fig. 2.2 (a) The score distribution of ALL Bordeaux Wines; (b) The number of wines that have been reviewed annually

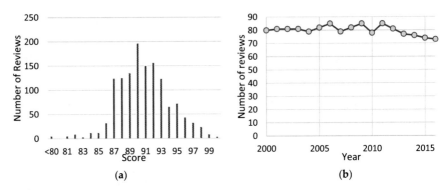

Fig. 2.3 (**a**) The score distribution of Bordeaux Wine Official Classification in 1855; (**b**) The number of wines reviewed annually

Classification in 1855 may still missing their wine reviews in Wine Spectator. A complete list of wines and vintages we cannot find within this dataset's scope is listed in Appendix B of [9].

References

1. Wine Spectator 100point scale. https://www.winespectator.com/articles/scoring-scale. Accessed 1 May 2022
2. Nobel, A.C.: Wine Aroma Wheel. http://winearomawheel.com/. Accessed 1 May 2022
3. Wine Aroma Wheel from UCA Davis. https://www.thewinecellarinsider.com/wine-topics/wine-educational-questions/davis-aroma-wheel/. Accessed 1 May 2022
4. Juluru, K., Shih, H.-H., Murthy, K.N.K., Elnajjar, P.: Bag-of-words technique in natural language processing: a primer for radiologists. Radiographics. **41**(5), 1420–1426 (2021)
5. Voorhees, E.M.: Natural language processing and information retrieval. In: International Summer School on Information Extraction, pp. 32–48. Springer, Berlin, Heidelberg (1999)
6. Mayfield, J., McNamee, P.: Single n-gram stemming. In: Proceedings of the 26th Annual International ACM SIGIR Conference on Research and Development in Information Retrieval, pp. 415–416 (2003)
7. Kondrak, G.: N-gram similarity and distance. In: International Symposium on String Processing and Information Retrieval, pp. 115–126. Springer, Berlin, Heidelberg (2005)
8. Chen, B., Rhodes, C., Crawford, A., Hambuchen, L.: Wineinformatics: applying data mining on wine sensory reviews processed by the computational wine wheel. In: 2014 IEEE International Conference on Data Mining Workshop, pp. 142–149. IEEE, Piscataway, NJ (2014)
9. Dong, Z., Guo, X., Rajana, S., Chen, B.: Understanding 21st century Bordeaux wines from wine reviews using naïve bayes classifier. Beverages. **6**(1), 5 (2020)

Chapter 3
Classification in Wineinformatics

Abstract *"How does a wine achieve 90+ scores?"* is the main problem to be solved in this chapter. In order to answer the question, the problem of classification with a target label, different types of classification algorithms and evaluation methods are presented. Classification models are built on all three datasets to predict whether a wine may receive a score above 90 points.

3.1 Classification and Classification Label

Supervised learning in machine learning uses the labeled training dataset to create a method that maps an input to an output. If the label in the training dataset is a continuous value, it is known as **Regression**; if the label in the training dataset is a discrete and unordered variable, it is known as **Classification** [1]. For example, regression in Wineinformatics predicts the actual grade of a wine, while classification in Wineinformatics predicts the categorical (good, very good, outstanding, classic) label of a wine. As one would expect, the label decides the direction of the supervised learning outcomes.

In this chapter, to answer the question "*How does a wine achieve 90+ scores?*", the wine score is chosen as the label: if a wine achieves the score of 90 or above, a **positive (+)** label is assigned; if a wine scores of 89 or below, a **negative (−)** label is assigned. This setup makes the supervised learning into a traditional bi-class classification problem. Table 3.1 demonstrates the idea of appending the grade label to the end of each wine data.

3.2 Classification Algorithms

There are numerous learning algorithms in the classification problems; two major category of classification algorithms can be considered as **Lazy learners** and **Eager Learners**: Lazy learners put aside training dataset and wait until testing dataset comes and search for most relevant reference points in the training dataset. Eager

Table 3.1 Wine data example with attributes and label

ID	DARK PLUM	RASPBERRY	REFINED	RICH	GRADE
Wine1	1	1	0	1	+
Wine2	1	0	1	1	+
Wine3	1	0	1	1	+
Wine4	0	1	1	1	+
Wine5	0	1	0	1	+
Wine6	1	1	0	0	+
Wine7	0	1	0	0	−
Wine8	0	1	0	1	−
Wine9	1	1	0	0	−
Wine10	0	1	1	0	−

learners build the model through training dataset before the testing dataset comes. Lazy learners usually take less time in training but more time in testing while eager learners take more time in training but less time in testing.

Among all eager learners, depending on whether the model can be explained why certain predictions are made to the stakeholder, black-box and white-box classification algorithms are distinguished. **Black-box** classification algorithms, such as neural networks [2] and SVM [3], transform the problem into complicated ensembles or high dimensional space often provide great prediction performances. However, black-box algorithms are usually hard to understand and uneasy to extract knowledge based on the built model without further analysis. It only examines the fundamental aspects of the system and has no or little relevance with the internal logical structure of the system" [4]. **White-box** classification algorithms, on the other side, creates a human explainable model for prediction and easier for knowledge extraction. Famous white-box classification algorithms include decision tree [5] and Bayesian networks [6]. However, these models may suffer from oversimplify the problem result in lower predication performances. There is no clear line between black-box and white-box algorithms, with proper interpretation techniques, black-box models can extract decision information with higher cost of computational complexity and execution time.

Due to space limitation and the performance based on the previous research results [7, 8], this book focuses on two major classification algorithms that stands out in Wineinformatics researches: **SVM**, a black-box classification algorithm and **Naïve Bayes**, a white-box classification algorithm.

3.2.1   Support Vector Machines (SVM)

Support Vector Machines (SVM) were developed over a period of several decades. In 1963, Vapnik and Lerner introduced a method of pattern recognition using

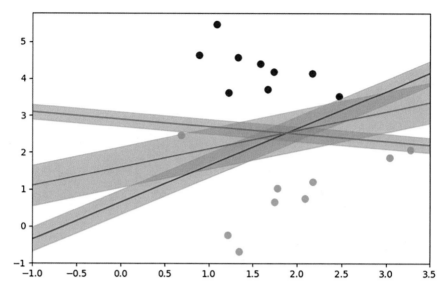

Fig. 3.1 SVM demonstrates the process of looking for a hyperplane/line with maximum margin [15]

generalized portraits, one of the first contributions specially towards inventing the SVM [9]. Even so, it was not until 1992 when Boser, Guyon, and Vapnik published a paper that advanced SVM close to their current form by introducing an algorithm for optimal margin classifiers [10]. Finally, the modern SVM arrived with the idea of the kernel trick, as well as support for regression, which was also added in the 1990s [11, 12]. Nowadays, the SVM are considered as having high utility to solve classification problems [13].

Each instance is plotted as a point in the n-dimensional space with their corresponding number of n-attributes in for SVM classification problems. After the n-dimensional space is defined, SVM looks for a hyperplane with maximum margin to separate different classes in the dataset, where margin is the distance between the hyperplane and the edge points from different classes [14]. The instances on the edge are called the "support vectors". Figure 3.1 shows how SVM work for a typical bi-class classification.

Consider blue and green dots represent two different classes, which may be separated by several different lines. However, the best choice will be the line that leaves the maximum margin from both classes. The black line in Fig. 3.2 will be the most preferable choice; also, the black points in the same figure are considered as "support vectors" for this sample data determining the orientation of the line [16]. While the ideal hard margin, a margin without points between the margin and separating hyperplane, is not available, the soft margin techniques allow errors to trade-off between maximizing the margin and minimizing the loss [17, 18]. SVM light [19] with linear kernel is the version of SVM that was used to perform the classification in this research. The linear kernel is one of the most commonly used

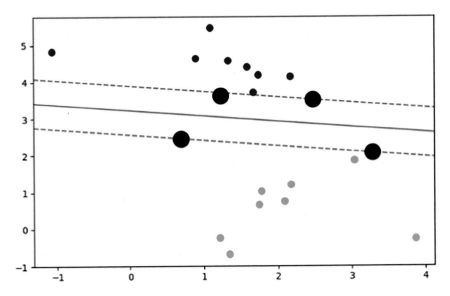

Fig. 3.2 The hyperplane/line with maximum margin in SVM [15]

kernels in practice, especially in text classification, where the dataset is usually in a high-dimensional, sparse feature space and linearly separable state [20, 21].

3.2.2 Naïve Bayes Classifier

Among all white-box classification algorithms applied on Wineinformatics, including decision trees, association classification, KNN, the Naïve Bayes classifier generates the best performance most of the time [7, 8]. A Naïve Bayes classifier is a simple probabilistic classifier by applying Bayes' theorem with two assumptions: First, there is **no dependence** between attributes. Second, in terms of the importance of the label/outcome, each attribute is treated **the same**. For example, the word DARK PLUM and the word REFINED have no influence with the appearance of each words and have equal importance in influencing the prediction of wine quality. In this instance, the algorithm presupposes that these two attributes are completely independent of each other, which is the basis for the name "Naïve."; because of that, Naïve Bayes classifier requires low computational complexity and is easy to apply on large amount of high dimensional data. Figure 3.3 provides the basic formula for Naïve Bayes classifier.

- Let **x** be a testing wine sample that the classifier wants to make the prediction; Let c be a hypothesis (our prediction) that x belongs to class C; Classification is to determine P(c|**x**), the probability that the hypothesis holds given the observed data

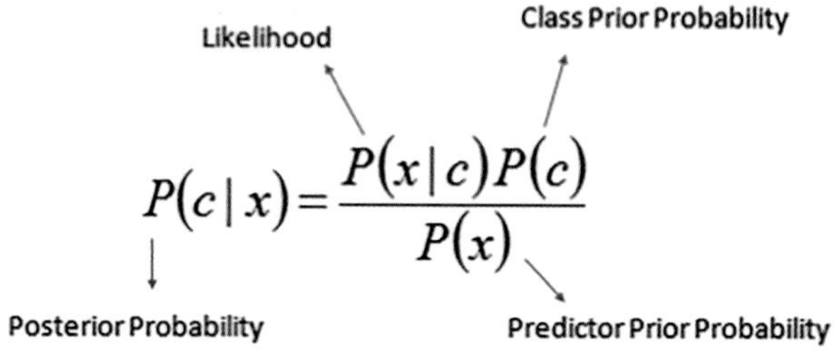

$$P(c \mid x) = \frac{P(x \mid c)P(c)}{P(x)}$$

Likelihood → numerator $P(x \mid c)$

Class Prior Probability → $P(c)$

Posterior Probability → $P(c \mid x)$

Predictor Prior Probability → $P(x)$

$$P(c \mid X) = P(x_1 \mid c) \times P(x_2 \mid c) \times \cdots \times P(x_n \mid c) \times P(c)$$

Fig. 3.3 Bayes' Theorem, the basis for the Naïve Bayes algorithm [22]

Table 3.2 Unknown wine grade data for prediction using Naïve Bayes Classifier

ID	DARK PLUM	RASPBERRY	REFINED	RICH	GRADE
Wine_x	1	0	0	1	?

sample **x**. For example, c can be the wine for testing and c can be the 90+ class, therefore, P(90 + |**x**) means the probability for testing wine belong to 90+ class.

- P(c) (*prior probability*), the initial probability for the class. For example, what is the original probably for a wine to receive 90+ class in the training dataset. Using wine data in Table 3.1 as the example, since it has six 90+ wines and four 89− wines; P(90+) = 60% and P(89−) = 40%
- P(**x**): probability that sample data is observed. In a traditional bi-class classification, the prediction is made by calculating the probability of two classes and the higher probability one wins. Since the x is the same testing dataset, P(x) usually does not need to be included in the calculation.
- P(**x**|c) (*posteriori probability*), the probability of observing the sample **x**, given that the hypothesis holds.
- P(**x**|c) = P(x1|c) * P(x2|c)*. . .P(xn|c) where x1, x2, . . . xn represent each attribute of x. In other words, P(x1|c) means what is the probability of attribute x1 occurs in each class in the training dataset.

Let us take a look of a concrete example, assume the Wine_x in Table 3.2, which has DARK PLUM and RICH ATTRIBUTES, is the testing data for prediction using Naïve Bayes classifier trained on Table 3.1's dataset. Here is how the Naïve Bayes classifier make the prediction:

- DARK PLUM attribute with positive occurs 4 times in 90+ wines and only once in 89− wines in the training dataset, so if the testing wine has positive DARK PLUM attribute, P(DARK PLUM|90+) = 4/6 and P(DARK PLUM|89−) = 1/4.

- RASPBERRY attribute with negative occurs 2 times in 90+ wines and 4 times in 89− wines in the training dataset, so if the testing wine has negative RASPBERRY attribute, P(RASPBERRY|90+) = 2/6 and P(RASPBERRY|89−) = 4/4.
- REFINED attribute with negative occurs 3 times in 90+ wines and 1 times in 89− wines in the training dataset, so if the testing wine has negative REFINED attribute, P(REFINED|90+) = 3/6 and P(REFINED|89−) = 1/4.
- RICH attribute with positive occurs 5 times in 90+ wines and only once in 89− wines in the training dataset, so if the testing wine has positive RICH attribute, P(RICH|90+) = 5/6 and P(RICH|89−) = 1/4.
- Plus, as mentioned above, P(90+) = 60% and P(89−) = 40%

Based on formula in Fig. 3.3:

- P(90+ |x) = (4/6) × (2/6) × (3/6) × (5/6) × (0.6) = 0.05555
- P(89− |x) = (1/4) × (4/4) × (1/4) × (1/4) × (0.4) = 0.00625

Since P(90+ |x) is greater than P(89− |x), the prediction to the unknown wine would be 90+ wines. Compare with SVM, a black-box method, Naïve Bayes demonstrate a clear logic for making predictions.

However, there is one catch: if Wine8 in Table 3.1 has **negative** in RICH attribute, which means no 89− wines in training dataset ever occurs positive in RICH attribute, P(RICH|89−) = 0/4. This incident will make P(89−|x) = 0 regardless all other attributes contribution nor p(89−) since anything multiply by 0 equals to 0. Therefore, another version of the Naïve Bayes algorithm is called **Laplacian smoothing**. Laplace smoothing alleviates this issue by adding a parameter such as one to both the numerator and the denominator so that these zero probabilities do not interfere with the probabilities of other attributes. Using the same example mentioned above, P(DARK PLUM |90+) = (4 + 1)/(6 + 1) and P(DARK PLUM |89−) = (1 + 1)/(4 + 1). This is important to note because we have made tests on our dataset with both the original Naïve Bayes algorithm as well as the Laplacian version.

3.3 Classification Evaluations

If the model can accurately predict a wine's grade category through the review, it means the model can capture the essence of high-quality wines. Therefore, before the classification results are demonstrated, a formal evaluation process and evaluation metrics need to be discussed.

Fig. 3.4 This figure demonstrates training and testing sets assigning in Fivefold cross-validation [15]

3.3.1 Fivefold Cross Validation

In this book, Fivefold cross-validation, illustrated in Fig. 3.4, is used as a process to evaluate the predictive performance of classification models.

First of all, shuffle the dataset randomly to avoid all positive or negative cases allocate on the same side. Then divide the dataset evenly into five portions. Finally, the program uses the subset 1 as the test set and the rest of subsets as the training set as fold 1; the program uses subset 2 as the test set, and the rest of the subsets as the training set which is set as fold 2; the program repeats the same process for the rest. In this case, each wine in the dataset has fair chances to become training and testing dataset. Each fold will have their evaluations and the final overall evaluation will be the average of each fold.

3.3.2 Evaluation Metrics

To evaluate the effectiveness of the classification model, several standard statistical evaluation metrics are used in this book. Firstly, four components, Positive (TP), True Negative (TN), False Positive (FP) and False Negative (FN), in the confusion matrix need to be defined. If the bi-class (90+ or 89−) classification problem trues to predict whether a wine scores above 90 points, we can view.

TP: The real condition is true (1) and predicted as true (1); 90+ wine correctly classified as 90+ wine;

TN: The real condition is false (−1) and predicted as false (−1); 89− wine correctly classified as 89− wine;

FP: The real condition is false (−1) but predicted as true (1); 89− wine incorrectly classified as 90+ wine;

FN: The real condition is true (1) but predicted as false (−1); 90+ wine incorrectly classified as 89− wine;

Once the components in the confusion matrix is defined, following evaluation metrics can now be introduced:

Accuracy: The proportion of wines that has been correctly classified among all wines. Accuracy is a very intuitive metric.

$$Accuracy = \frac{TP + TN}{TP + TN + FP + FN}$$

Precision: The proportion of predicted 90+ wines was actually correct.

$$Precision = \frac{TP}{TP + FP}$$

Recall/Sensitivity: The proportion of 90+ wines among all 90+ wines was identified correctly.

$$Recall/Sensitivity = \frac{TP}{TP + FN}$$

Specificity: The proportion of 89− wines among all 89− wines was identified correctly.

$$Specificity = \frac{TN}{TN + FP}$$

F-score: The harmonic mean of recall and precision. F-score takes both recall and precision into account, combining them into a single metric.

$$F\text{-}score = 2 \times \frac{precision * recall}{(precision + recall)}$$

3.4   How Does a Wine Achieve 90+ Scores?

To understand how does a wine achieve 90+ scores, all three datasets were labeled by 90+ or 89− scores, applied by both SVM and Naïve Byes classifier and evaluated through fivefold cross validation methods.

3.4.1   Dataset 1: The Big Dataset Results

In the big datasets, both Naïve Bayes Laplace classifier and SVM classifier achieved 83.9% accuracy or above. Details results can be found in Table 3.3. SVM classifier achieved the higher accuracy of 86.35%. In terms of precision, SVM classifier had a much better performance than Naïve Bayes with Laplace, which indicates that SVM has a lower false-positive rate than SVM classifier; In terms of recall, both classifier have similar results with less than 0.5% differences; Since SVM has better precision

Table 3.3 Accuracy, precision, recall, and F-score in different classifiers on Dataset1

	Accuracy	Precision	Recall	F-Score
Naïve Bayes Laplace	83.93%	78.38%	73.84%	76.04%
SVM	**86.35%**	**84.03%**	**74.32%**	**78.88%**

Table 3.4 Accuracy, precision, recall, and F-score in different classifiers on Dataset2

	Accuracy	Precision	Recall	F-Score
Naïve Bayes Laplace	85.17%	73.22%	**79.03%**	76.01%
SVM	**86.97%**	**80.68%**	73.80%	**77.10%**

Table 3.5 Accuracy, precision, recall, and F-score in different classifiers on Dataset3

	Accuracy	Precision	Recall	F-Score
Naïve Bayes Laplace	**84.62%**	86.79%	**90.02%**	**88.38%**
SVM	81.38%	**86.84%**	84.12%	85.46%

and recall, it also has a higher F-score. In this dataset, SVM has outperformed Naive Bayes classifier in all measurements.

3.4.2 Dataset 2: Twenty-First Century Bordeaux Dataset

In twenty-first century Bordeaux wine datasets, both Naïve Bayes Laplace classifier and SVM classifier achieved 85% accuracy or above. Details can be found in Table 3.4. SVM classifier achieved the highest accuracy of 86.97%. In terms of precision, SVM classifier had a much better performance than Naïve Bayes Laplace classifier, which indicates that SVM classifier has a lower false-positive rate than Naïve Bayes Laplace classifier. Diametrically opposed to its recall, Naïve Bayes Laplace classifier had a much better performance, which indicates that Naïve Bayes Laplace classifier had a lower false negative rate than SVM classifier. Naïve Bayes classifier and SVM classifier have very similar F-scores, but SVM classifier is slightly better. Overall, SVM has a slightly better performance in terms of accuracy and F-score in this dataset.

3.4.3 Dataset 3: Twenty-First Century Elite Bordeaux

In the twenty-first century elite Bordeaux dataset, both the Naïve Bayes Laplace classifier and SVM classifier are able to achieve 81% accuracy or above. Details can be found in Table 3.5. Unlike previous two datasets, this one has more positive cases (90+) than negative cases (89−). Naive Bayes Laplace classifier achieves the best accuracy of 84.62%. In terms of precision, Naïve Bayes Laplace classifier and SVM classifier all achieve around 86%. In terms of recall, Naive Bayes Laplace classifier achieves the best recall of 90.02%, which is around 6% higher than SVM classifier. Naïve Bayes Laplace classifier has much better sensitivity than SVM classifier. In the combination of precision and recall, Naive Bayes Laplace classifier has the highest F-score of 88.38%. Overall, Naïve Bayes Laplace had a better performance than the SVM classifier in this specific elite Bordeaux wine dataset.

In summary, all three dataset achieves more than 80% accuracy when predicting whether a wine achieves 90 points or more. This satisfactory results suggest that the both SVM and Naïve Bayes classifier are able to create a valid model to capture reviewers' wine evaluation logic. This also suggest that the process of wine reviews collection, natural language processing model creation has the potential to provide useful knowledge in Wineinformatics.

References

1. Kotsiantis, S.B., Zaharakis, I., Pintelas, P.: Supervised machine learning: a review of classification techniques. In: Emerging Artificial Intelligence Applications in Computer Engineering, vol. 160, pp. 3–24 (2007)
2. Abiodun, O.I., Jantan, A., Omolara, A.E., Dada, K.V., Mohamed, N.A.E., Arshad, H.: State-of-the-art in artificial neural network applications: a survey. Heliyon. 4(11), e00938 (2018)
3. Hearst, M.A., Dumais, S.T., Osuna, E., Platt, J., Scholkopf, B.: Support vector machines. In: IEEE Intelligent Systems and Their Applications, vol. 13, pp. 18–28 (1998)
4. Khan, M., Khan, F.: A comparative study of White Box, Black Box and Grey Box Testing Techniques. IJACRA. 3(6), 12 (2012)
5. Myles, A.J., Feudale, R.N., Liu, Y., Woody, N.A., Brown, S.D.: An introduction to decision tree modeling. J. Chemometrics. 18(6), 275–285 (2004)
6. Friedman, N., Geiger, D., Goldszmidt, M.: Bayesian network classifiers. Mach. Learn. 29(2), 131–163 (1997)
7. Chen, B., Le, H., Atkison, T., Che, D.: A Wineinformatics study for white-box classification algorithms to understand and evaluate wine judges. Trans. Mach. Learn. Data Min. 10(1), 3–24 (2017)
8. Le, H.: Classification on Wine Informatics. University of Central Arkansas (2015)
9. Vapnik, V.N.: Pattern recognition using generalized portraits [translated from Vapnik and Lerner (1963b)]. Autom. Remote. Control. 24(6), 709–715 (1963)
10. Boser, B.E., Guyon, I.M., Vapnik, V.N.: A training algorithm for optimal margin classifiers. In: Proceedings of the Fifth Annual Workshop on Computational Learning Theory, pp. 144–152 (1992)
11. Poggio, T., Girosi, F.: Networks for approximation and learning. Proc. IEEE. 78(9), 1481–1497 (1990)
12. Cortes, C., Vapnik, V.: Support-vector networks. Mach. Learn. 20(3), 273–297 (1995)
13. Fernández-Delgado, M., Cernadas, E., Barro, S., Amorim, D.: Do we need hundreds of classifiers to solve real world classification problems? J. Mach. Learn. Res. 15(1), 3133–3181 (2014)
14. Suykens, K.J.A., Vandewalle, J.: Least squares support vector machine classifiers. Neural Process. Lett. 9, 293–300 (1999)
15. Dong, Z.: Understanding 21st Century Bordeaux Wines from Wine Reviews Through Natural Language Processing and Classifications. University of Central Arkansas (2020)

16. Favorov, O., Macdonald, J., Kursun, O.: SVM-based analysis of NMR spectra in metabolomics: development of procedures. J. Sci. Med. **1**(2) (2019)
17. Wu, Q., Zhou, D.-X.: SVM soft margin classifiers: linear programming versus quadratic programming. Neural Comput. **17**(5), 1160–1187 (2005)
18. Chen, D.-R., Qiang, W., Ying, Y., Zhou, D.-X.: Support vector machine soft margin classifiers: error analysis. J. Mach. Learn. Res. **5**, 1143–1175 (2004)
19. Thorsten, J.: Svmlight: support vector machine. https://www.researchgate.net/profile/Thorsten_Joachims/publication/243763293_SVMLight_Support_Vector_Machine/links/5b0eb5c2a6fdcc80995ac3d5/SVMLight-Support-Vector-Machine.pdf
20. Leopold, E., Kindermann, J.: Text categorization with support vector machines how to represent texts in input space? Mach. Learn. **46**(1–3), 423–444 (2002)
21. Joachims, T.: Text categorization with support vector machines: learning with many relevant features. In: Machine Learning: ECML-98. Lecture Notes in Computer Science, pp. 137–142 (1998)
22. Velchev, V.D.: Wineinformatics: a quantitative analysis of wine reviewers. Master thesis, University of Central Arkansas, Department of Computer Science (2017)

Chapter 4
Evaluation of Wine Judges

Abstract *"Are wine reviewers reliable and consistent?"* is always a big problem in wine review process. This chapter investigates the consistency of the wine being considered as "outstanding" or "extraordinary" using the dataset1: The Big dataset.

4.1 Wine Reviewers in Wine Spectator

"Are wine reviewers reliable and consistent?" is always a big problem in wine review process. According to the Journal of Wine Economics [1], questions such as, "Who is a reliable wine judge? Are wine judges consistent? Do wine judges agree with each other?" are required for formal statistical answer. In the past decade, many researchers focused on these problems with small to medium sized wine datasets [2–5]. However, no research is being performed on analyzing the consistency of wine judges with a large-scale dataset. As a prestigious magazine in the wine field, WineSpectator.com contains more than 370,000 wine reviews. This chapter investigates the consistency of the wine being considered as "outstanding" or "extraordinary" using the dataset1: The Big dataset. Wine Spectator reviews more than 15,000 wines per year and these reviews comes from 10 reviewers. Table 4.1 shows the position of each reviewer and the "tasting beat" or wine regions from which they taste.

Table 4.2 displays the score distribution for the reviewers in dataset1: The Big dataset. As mentioned in Chap. 2, more than 50% of the wine falls in to the "Very Good" category and only 1.5% of the wine receive the honor of "Classic". Compare with all other nine reviewers Gillian Sciaretta (GS) reviewed much less wines; Due to the lacking sample size of Gillian Sciaretta, specifically in the top two categories, in this chapter, she was excluded from the evaluation of Wine Spectator reviewers.

B. Chen, *Wineinformatics*, SpringerBriefs in Computer Science,
https://doi.org/10.1007/978-981-19-7369-7_4

Table 4.1 Wine Spectator reviewer profiles

Reviewer	Position	Tasting Beat
James Laube (JL)	Senior editor, Napa	California
Kim Marcus (KM)	Managing editor, New York	Argentina, Austria, Chile, Germany, Portugal
Thomas Mat- thews (TM)	Executive editor, New York	New York, Spain
James Molesworth (JM)	Senior editor, New York	Bordeaux, Finger Lakes, Loire Valley, Rhône Valley, South Africa
Bruce Sanderson (BS)	Senior editor, New York	Burgundy, Italy
Harvey Steiman (HS)	Editor at large, San Francisco	Australia, Oregon, Washington
Tim Fish (TF)	Senior editor, Napa	California Merlot, Zinfandel and Rhône-style wines, U.S. sparkling wines
Alison Napjus (AN)	Senior editor and tasting director, New York	Alsace, Beaujolais, Champagne, Italy
MaryAnn Worobiec (MW)	Senior editor and senior tasting coordinator, Napa	Australia, California (Petite Sirah, Sauvignon Blanc, other whites) and New Zealand
Gillian Sciaretta (GS)	Tasting coordinator, New York	France

Table 4.2 Wine Spectator review metadata from 2006–2015

Reviewer Judge	80–84 (Good)	85–89 (Very Good)	90–94 (Outstanding)	95–100 (Classic)
James Laube (JL)	1250	7384	5168	357
Kim Marcus (KM)	1618	6690	3217	161
Thomas Matthews (TM)	1367	2981	1144	26
James Molesworth (JM)	3857	13,628	6682	433
Bruce Sanderson (BS)	1148	7677	8618	451
Harvey Steiman (HS)	708	7657	5755	178
Tim Fish (TF)	1236	2531	1032	16
Alison Napjus (AN)	1510	4802	2095	33
MaryAnn Worobiec (MW)	833	3745	676	15
Gillian Sciaretta (GS)	66	355	7	0
Total	13,593	57,450	34,394	1670

4.2 How to Evaluate Wine Reviewers

Wine reviews provided by the wine judges are the fundamental component in Wineinformatics research. These reviews come from wine judges taste wines to determine which attributes the wine exhibits, describe them in words and give a verdict to the wine.

The question this chapter seeks to focus on is the consistency of the wine judges' evaluation. If a wine reviewer does not agree with his/her own review on the same or similar wine, computation models will not be able to create a valid logic for prediction. More specifically, when a wine judge gives a wine a 90+ score with the wine review as the evaluation; does the wine judge give a 90+ score to other wines with identical or very similar reviews? Based on this question, the wine dataset is labeled into two categories: the wines that score 90+ and the wines that score 89−, which is the same labeling method used in the previous chapter.

The logic of classification algorithms is to use a collection of previously categorized data as a basis for categorizing new data. Therefore, the data will include a training set from which to base its classification on as well as a testing set that will predict unknown quantities into the previously established categories. Having more consistent wine reviews from each wine reviewer will lead to a more accurate classification model which will yield better evaluation results for the testing dataset.

For each reviewer, their reviews included in the Dataset1: The Big Dataset are individually collected (summarized in Table 4.2); for example, dataset from James Laube (JL) includes 14,159 wines (1250 + 7384 + 5168 + 357). Therefore, **nine** datasets (with GS dataset excluded) in total are generated and evaluated through both SVM and Naïve Bayes Classifier using fivefold cross validation. The performance of each dataset represents the consistency of the corresponding wine judge.

4.3 Ranking of the Wine Reviewers Using Naïve Bayes Classifier

The complete evaluation results for each wine reviewer based on Naïve Bayes algorithm are given in Table 4.3. Both original Naïve Bayes and Laplace smoothing results are included. The Naïve Bayes Original algorithm performed slightly worse than the Laplace. **1**, as the parameter, was added to both numerator and the denominator to all P(xn|c) calculation so that these zero probabilities do not interfere with the probabilities of other attributes.

The most reliable reviewer in this instance was Tim Fish (TF), who had an accuracy of 87.37% with the original version of the algorithm and 88.16% with the Laplace version. However, MaryAnn Worobiec (MW) was a very close second with 87.36% and 88.04% respectively. All of the reviewers had a very high precision, generally around 80%, with the exception of MaryAnn Worobiec (MW), one of the two higher scoring reviewers. Mining this type of information can give insight

Table 4.3 Complete Naïve Bayes Classifier Results [6]

Reviewer: AN	Naïve Bayes original	Laplace correction
Accuracy	0.872867299	0.878909953
Precision	0.776315789	0.782424812
Recall/Sensitivity	0.734548688	0.748651079
Specificity	0.923114198	0.925514801
Reviewer: BS	Naïve Bayes original	Laplace correction
Accuracy	0.802615402	0.804794903
Precision	0.808247877	0.807034954
Recall/Sensitivity	0.803463773	0.807658354
Specificity	0.801732984	0.801856884
Reviewer: HS	Naïve Bayes original	Laplace correction
Accuracy	0.791019723	0.793817317
Precision	0.75526715	0.751559076
Recall/Sensitivity	0.744723284	0.751559076
Specificity	0.824658858	0.8237896
Reviewer: JL	Naïve Bayes original	Laplace correction
Accuracy	0.802810933	0.805918497
Precision	0.753303167	0.751674208
Recall/Sensitivity	0.74441066	0.751130403
Specificity	0.840919701	0.841019699
Reviewer: JM	Naïve Bayes original	Laplace correction
Accuracy	0.868211382	0.870406504
Precision	0.823471539	0.823893183
Recall/Sensitivity	0.746845124	0.751827626
Specificity	0.925037302	0.925429983
Reviewer: KM	Naïve Bayes original	Laplace correction
Accuracy	0.845284956	0.849478008
Precision	0.772646536	0.772350503
Recall/Sensitivity	0.715068493	0.72492359
Specificity	0.904430065	0.904909113
Reviewer: MW	Naïve Bayes original	Laplace correction
Accuracy	0.873600304	0.88043272
Precision	0.652677279	0.687409551
Recall/Sensitivity	0.514253136	0.534308211
Specificity	0.945355191	0.950684932
Reviewer: TF	Naïve Bayes original	Laplace correction
Accuracy	0.873727934	0.881619938
Precision	0.81870229	0.836832061
Recall/Sensitivity	0.672413793	0.687304075
Specificity	0.946312518	0.951681266
Reviewer: TM	Naïve Bayes original	Laplace correction
Accuracy	0.848314607	0.859188112
Precision	0.747863248	0.767521368

(continued)

Table 4.3 (continued)

Recall/Sensitivity	0.617501764	0.640057021
Specificity	0.928066325	0.933900365
AVERAGE	Naïve Bayes original	Laplace correction
Accuracy	0.842050282	0.847173995
Precision	0.767610542	0.775633302
Recall/Sensitivity	0.699247635	0.710824382
Specificity	0.893291905	0.895420738

Table 4.4 Reviewers by order of Naïve Bayes accuracy [6]

Reviewer	Original Naïve Bayes	Laplace
TF	87.37%	88.16%
MW	87.36%	88.04%
AN	87.28%	87.89%
JM	86.82%	87.04%
TM	84.83%	85.91%
KM	84.52%	84.94%
Average	84.20%	84.71%
JL	80.28%	80.59%
BS	80.26%	80.47%
HS	79.10%	79.38%

into which reviewers give precise descriptions in their reviews and with enough data collected on the reviewers, one could rank them by their reliability. In this case, by order of accuracy as shown in Table 4.4, these would be TF, MW, AN, JM, TM, KM, JL, BS, HS. Although HS has the lowest accuracy, he is only less than 1% lower than 80% accuracy with both Naïve Bayes classifiers. The average show in the table is the average of nine reviewers without considering the number of reviews they produced. If considered, the overall accuracy of the whole dataset using Naïve Bayes with Laplace is 83.93%, as shown in Chap. 3. To the best of our knowledge, this is the very first research work to rank different judges based on their large amount of reviews through data science methods.

Due to the skew of these datasets, with the vast majority of wines being below 90, or in the 89− category, experimental results generally reflected high specificity, mediocre precision and low recall for all reviewers. However, reviewer Bruce Sanderson (BS) remained the most consistent for our predictions despite the skew, with precision, recall and specificity all within one percentage point for both the original Naïve Bayes and the Laplace correction.

One major advantage of using a white-box classification algorithm is the logic of prediction can be understood. Table 4.5 shows the attribute that are heavily correlates to the positive label (90+ scores) at least 90% of the time with at least 30 instances of the attribute. These words represent preferable attributes used by each reviewer.

Based on Table 4.5, several reviewers have more positively correlated attributes in their 90+ wine reviews; it indicates that certain reviewers have words that they are

Table 4.5 Naïve Bayes Positively Correlated Attributes

Reviewer	Attributes correlated positively (>90 rating) with at least 30 instances
AN	Intense 30/33, Beauty 55/58, Power 57/59, Seamless 43/44, Finesse 41/45
BS	Alluring 103/112, Excellent 182/184, Terrific 170/175, Refined 171/182, Seamless 77/80, Potential 141/149, Detailed 104/114, Beauty 285/290, Seductive 35/37, Gorgeous 33/34, Ethereal 50/53
HS	Deep 58/61, Elegant 276/305, Power 156/172, Long 765/849, Impresses 214/232, Complex 238/264, Seductive 83/92, Beauty 215/228, Tension 33/36, Remarkable 29/32, Gorgeous 71/73, Tremendous 53/53
JL	Plush 93/101, Seductive 82/88, Delicious 97/104, Wonderful 125/130, Opulent 41/44, Beauty 114/115, Remarkable 33/33, Gorgeous 65/65, Amazing 57/57
JM	Rock Solid 155/171, Seamless 146/162, Impresses 186/192, Turkish Coffee 69/75, Packed 150/159, Serious 94/98, Remarkable 51/52, Gorgeous 360/361, Terrific 104/104, Beauty 148/148, Wonderful 49/49, Backward 36/36, Stunning 57/57
KM	Complex 173/188
TM	Long 72/78

likely to fall back on when describing quality wines. This information might be useful to make more accurate predictions about those particular reviewers and perhaps even their biases. Unsurprisingly, nearly all of the attributes are subjective adjectives used by the taste to describe the overall wine, such as "beauty" and "Elegant". However, some attribute such as "Turkish Coffee" may not have a high rating in general but only for some specific wine region or some particular reviewers. The attributes listed in the table may be used for creating even more powerful prediction models for a designated wine regions or reviewers.

4.4   Ranking of the Wine Reviewers Using SVM

The complete evaluation results for each wine reviewer based on SVM are given in Table 4.6.

Figure 4.1 is derived from Table 4.6 and shows all four evaluation metrics for each reviewer. Most of the reviews have very high specificity, high accuracy, low precision and very low recall except Bruce Sanderson (BS) whose reviews were evenly balanced between 90+ and 89− wines. Most of the reviewers resemble Alison Napjus's (AN) curve; some of the more exaggerated versions of this curve, such as MaryAnn Worobiec (MW) in the figure, happen when the reviewer rates their majority wine below 90 points (in her instance, 86.9% of her ratings are 89−).

Table 4.7 demonstrates the reviewers' rankings by SVM accuracy. According to the results generated by SVM, the most consistent reviewer is MaryAnn Worobiec because her reviews consistently have 91% accuracy. Tim Fish also holds higher than 90% accuracy as the second most consistent reviewer. James Molesworth, Thomas Mathews and Alison Napjus are the third, fourth and fifth ranked reviewers who holds higher than 88% accuracy, respectively. The average show in the table is

Table 4.6 Evaluation results for each wine reviewer based on SVM

	SVM performances
Reviewer: AN	
Accuracy	0.8836
Precision	0.8408
Recall/Sensitivity	0.7053
Specificity	0.9597
Reviewer: BS	
Accuracy	0.8312
Precision	0.8380
Recall/Sensitivity	0.8274
Specificity	0.8358
Reviewer: HS	
Accuracy	0.8264
Precision	0.8255
Recall/Sensitivity	0.7456
Specificity	0.8955
Reviewer: JL	
Accuracy	0.8247
Precision	0.8026
Recall/Sensitivity	0.7337
Specificity	0.8876
Reviewer: JM	
Accuracy	0.8935
Precision	0.8547
Recall/Sensitivity	0.7780
Specificity	0.9496
Reviewer: KM	
Accuracy	0.8729
Precision	0.8242
Recall/Sensitivity	0.7279
Specificity	0.9417
Reviewer: MW	
Accuracy	0.9138
Precision	0.9214
Recall/Sensitivity	0.5209
Specificity	0.9975
Reviewer: TF	
Accuracy	0.9121
Precision	0.8807
Recall/Sensitivity	0.7185
Specificity	0.9771
Reviewer: TM	
Accuracy	0.8905

(continued)

Table 4.6 (continued)

	SVM performances
Precision	0.8405
Recall/Sensitivity	0.6589
Specificity	0.9721
Average	
Accuracy	0.8721
Precision	0.8476
Recall/Sensitivity	0.7129
Specificity	0.9352

the average of nine reviewers without considering the number of reviews they produced. If considered, the overall accuracy of the whole dataset using SVM is 86.35%, as shown in Chap. 3.

4.5   Comparison of Naïve Bayes and SVM

The SVM algorithm had the most successful results but we cannot trace how it arrives at these results, due to the black-box nature of the algorithm. Reviewers MW and TF had the most successful results with the SVM as they did with the Naïve Bayes algorithm, both with more than 91% accuracy on their best performances. In order to compare the results obtained from two different classification methods, Table 4.8 is summarized from Tables 4.3, 4.4, 4.6, and 4.7. SVM, the black box classification algorithm, produced higher accuracy results as expected.

The average accuracy found in SVM reaches 87.21%; compare with the average accuracy found in Naïve Bayes (84.72%), SVM performed around 2.5% more accurate in average. All reviewers have better accuracy results; MW showed the biggest difference (improve 3.34%) while AN showed the least changes (improve 0.47%).

In terms of ranking of the reviewers, both methods give similar results. MW and TF are ranked 1st and 2nd in both SVM and Naïve Bayes. BS, HS and JL are ranked 7th, 8th and 9th in both classification methods. The middle rank 3rd, 4th and 5th show some minor differences. SVM ranked JM, TM and AN 3rd, 4th and 5th; while Naïve Bayes ranked AN, JM and TM 3rd, 4th and 5th. However, the accuracy between the three reviewers is very little. The results demonstrate in Table 4.8 suggest both methods have the capability to rank reviewers and be able to capture information between the reviews and the wines' score.

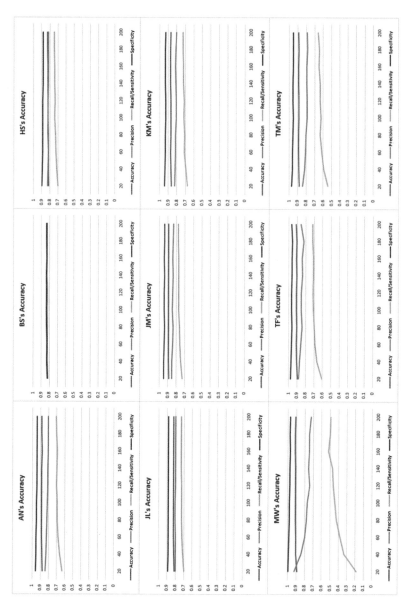

Fig. 4.1 Evaluation results for each wine reviewer based on SVM

Table 4.7 Reviewers by order of SVM accuracy

Reviewer	SVM peak
MW	0.9138
TF	0.9121
JM	0.8935
TM	0.8905
AN	0.8836
KM	0.8729
Average	0.8721
BS	0.8312
HS	0.8264
JL	0.8247

Table 4.8 Comparison results obtained from SVM and Naïve Bayes

Reviewer: AN	SVM	SVM Rank	Naïve Bayes with Laplace	Naïve Bayes Rank
AN	0.8836	5	0.8789	3
BS	0.8312	7	0.8047	8
HS	0.8264	8	0.7938	9
JL	0.8247	9	0.8059	7
JM	0.8935	3	0.8704	4
KM	0.8729	6	0.8494	6
MW	0.9138	1	0.8804	2
TF	0.9121	2	0.8816	1
TM	0.8905	4	0.8591	5
Average	0.8721	–	0.8471	–

References

1. Storchmann, K.: Introduction to the issue. J. Wine Econ. **10**, 1–3 (2015)
2. Quandt, R.E.: A note on a test for the sum of ranksums. J. Wine Econ. **2**, 98–102 (2007)
3. Ashton, R.H.: Improving experts' wine quality judgments: Two heads are better than one. J. Wine Econ. **6**, 135–159 (2011)
4. Ashton, R.H.: Reliability and consensus of experienced wine judges: Expertise within and between? J. Wine Econ. **7**, 70–87 (2012)
5. Bodington, J.C.: Evaluating wine-tasting results and randomness with a mixture of rank preference models. J. Wine Econ. **10**, 31–46 (2015)
6. Chen, B., Velchev, V., Palmer, J., Atkison, T.: Wineinformatics: A quantitative analysis of wine reviewers. Fermentation. **4**(4), 82 (2018)

Chapter 5
Regression in Wineinformatics

Abstract Regression analyses are applied on the big dataset to work on the question of "**Can actual wine grade and price be predicted through their reviews?**" This chapter provides an opportunity to discover the relationship between the exact wine grade and the combination of flavor, wine body and non-flavor descriptions. This chapter also seeks to reveal the suitable price of the wine from wine reviewers' perspective instead of wine makers' decisions. Several new evaluation metrics for regression are introduced and used for model evaluation. The results of this chapter actually demonstrate the challenge of predicting the cost of a wine based on its reviews.

5.1 Regression Analysis Data

There are two major factors reflecting the quality of the wine, one is the grapes itself, which are affected by the grape growing process, such as terroir and weather; while another one is the wine making process, which is governed by the wine makers. Experienced wine-makers can taste the grape harvests and build up the optimal wine-making process, which contains the following five stages: harvesting, crushing and pressing, fermentation, clarification, and aging and bottling. During the "crushing and pressing" process, the flavor, color, and tannins of the wine are defined; during the "fermentation" process, the acidity, alcohol percentage, and sweetness of the wine are carefully monitored; and during the "aging and bottling" process, the final touches of the wine are overseen, such as aging in French oak barrels for a subtle, spicy, and silky texture, while aging in American oak barrels will yield a stronger flavor, such as in cream soda, vanilla, and coconut [1].

The dataset used for this chapter is still Dataset1: The Big Dataset. The generic regression model build from wines from the world are expected to predict both wine grades and prices instead of the class of the wine. The quality of the wine was judged based on Wine Spectator's 100-point scale. Prices ranged from $0 to almost $1000 per 750 mL bottle of wine. The boxplots of the grades and prices are shown in Figs. 5.1 and 5.2, respectively. In a boxplot, Inter Quartile Range (IQR) is defined as 25th to the 75th percentile, which tells how spread the middle values are. Everything

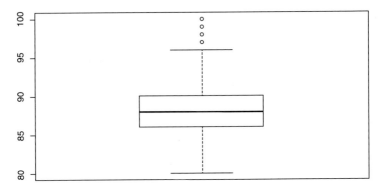

Fig. 5.1 Boxplot of grade with all wines [1]

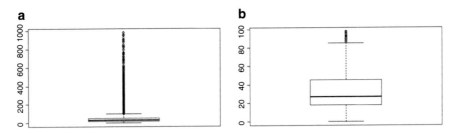

Fig. 5.2 Boxplot of prices with all wines (**a**) and boxplot of prices without outliers (**b**) [1]

outside of maximum, 75th percentile + 1.5*IQR, or minimum, 25th percentile − 1.5*IQR, are considered as outliers. IQR is drawn as the box and outliers are drawn as dots in the boxplot.

In terms of the prices, there are MANY extreme values compare with the average wine price per bottle. For example, Chateau Latour 2009 costs more than $1400 per bottle; while Chateau Petrus 2015 costs around $4000 per bottle; while the average wine price is close to $30. These extreme outliers may cause huge errors to the regression model; therefore, a separate dataset that all outliers showed in Fig. 5.2a are removed form wine price regression model were created for comparison purposes. This extra step removes the influence of overly expensive wines to the regression model. Figure 5.2b shows the boxplot for the updated dataset.

5.2 Methods

5.2.1 Support Vector Regression (SVR)

Support Vector Regression (SVR) [2–4] which is based on SVM, was the choice of algorithm for this chapter. SVM draw a hyperplane between two classes of data and maximize the margin to perform classification; while SVR draw a hyperplane that follows a trend in the data and maximized the number of data fit into the margin. SVR has two important parameters for the hyperplane: the kernel (linear and RBF were applied in this work) and epsilon, which control the slack in the margins. Allowing for the slack in the margins allows for dealing with the case when not all data fits within the margins around the hyperplane.

The implementation of SVR used the Kernlab package in R, with Microsoft's R Open [5–7]. The gof function from hydroGOF was also used perform the evaluations.

5.2.2 Evaluation Metrics

Regression model can be evaluated in many different metrics. Based on the testing data's attributes, regression model predicts an actual number of the target, such as a grade or a price. The difference between the predicted value and the actual value is so called "error". If the predicted value is higher than the actual value, the error is positive; If the predicted value is lower than the actual value, the error is negative. The average of all errors in the testing dataset is known as the **mean error (ME)**.

However, the mean error (ME) equals to zero does not mean the model makes a perfect prediction since the positive errors and negative errors may cancel each other when an average value is taken. Therefore, **mean absolute error (MAE)** is a better metric; MAE calculates how much error accumulated on average from the entire prediction process. Computing this metric is the same as with ME, except one simply takes the absolute value of the error before adding it into the average.

Another metric is the **mean squared error (MSE)**, which squares the error value and take the average on the all accumulated squared errors during the prediction process. MSE is better in detecting a few but large errors are made but it may also sensitive to outliers. Since simply squaring the errors results in enormous values, it is common to take the square root after the average is done, resulting in a **root mean squared error (RMSE)**. This also transforms the error back into the same units as the value being predicted.

Both MAE and RMSE give a real number which reflects the actual value of error; yet the number is not normalized into the actual range of predicting values. For example, while a regression model performed a prediction with MAE = 1; without knowing the possible range of targeting value (it can be 0–100 points or 0–10,000 $), it is hard to judge the result is good or bad. If the range is embedded to the evaluation metric, the normalized values allow better understanding of the performances. There are two normalize methods used in this work: The first way is to take the error and divide it by the range of possible values. This can be expressed by the formula:

$$Minimax\ Norm\ Error_y = \frac{Error_y}{\max\ (y) - \min\ (y)} \times 100$$

where y is the variable being predicted. Multiplying by 100 transforms the relative error into an easy-to-interpret percentage value. Normalizing this way makes sense when clear boundaries can be established for the maximum and minimum values, which is the case for the grade—the minimum and maximum grades in the Dataset1, The Big Dataset are 80 and 100. However, the price prediction does not fit this matric very well simply because the maximum price can be extreme. Therefore, a more reliable normalized method that can mitigates the impacts on outliers is using the standard deviation. Rather than taking the error and dividing it by the range of the data, dividing the error by the standard deviation makes the normalized error more robust to outliers and removes the need for us to know what the range of prices could be. This formula is:

$$Standard\ Deviation\ Norm\ Error_y = \frac{Error_y}{\sigma_y} \times 100$$

Where y is the variable being predicted and σ is the standard deviation of the variable.

In the results section, grades prediction will use the minimax normalization and prices prediction will use the standard deviation normalization. The normalization will be applied to MAE and RMSE so the notation will become **NMAE** and **NRMSE**, where N stands for normalized.

5.3   Regression Analysis

5.3.1   Can Wine Grade Be Predicted Through Their Reviews?

Table 5.1 shows the regression results for running SVR on dataset1 for grade prediction with both kernels.

Close to zero ME value in the table for all kernels suggests that the SVR is able to find a well-balanced hyperplane. MAE close to 1.6 for all kernels indicates the SVR is able to predict the actual grade within 2 points of error; RMSE close to 2 also suggests the same results. MSE close to 4 indicates there are no large error; rather, there are many smaller errors, as expected with a good regression model. The

Table 5.1 Table of results for the regression model on Grade. Normalized error is calculated using the minimax method

Kernel	ME	MAE	MSE	RMSE	NMAE	NRMSE
Linear	−0.07	1.64	4.27	2.07	8.2%	10.4%
RBF	0.01	1.59	4.02	2.00	8.0%	10.0%

normalized errors were also smaller, suggesting that the model can predict within about 8% (1.6 points) of the actual value of the instance. The satisfactory results suggest that there is a strong correlation between wine grade with wine reviews processed by the Computational Wine Wheel.

5.3.2 Can Wine Price Be Predicted Through Their Reviews?

Table 5.2 shows the regression results for running SVR on dataset1 for price prediction with three different kernels. Since the dataset uses the whole dataset1, a clear evidence of outliers can be observed: extremely large MSE value and the RMSE is over twice the value of MAR, indicates that there are several large errors. In addition, the ME was very negative, suggesting that the actual values were higher than the model's predictions. Again, the normalized method used in this subsection is through standard deviation, thus, the normalized errors are not comparable to the normalized errors for grade. Over 40% of NMAE indicates that the predicted price is almost half a standard deviation away from the actual value; over 90% of NRMSE also shows that the predicted price, after squaring and rooting the error, is almost a whole standard deviation away from the actual value. Obviously, there's room for improvement.

Considering the distribution of prices, according to Fig. 5.2a, there are many outliers, which caused many large errors to regression evaluation metrics. Based on the boxplot, exactly 6800 wines with prices higher $98, which is the maximum price to be considered as normal, or non-outlier. By removing 6800 from dataset1 that contains more than 107,000 wine, the boxplot becomes much less dominated by outliers, as shown in Fig. 5.2b. The new dataset derived from dataset1 allows SVR regression model to perform more precise predictions. Table 5.3 shows the regression results for running SVR on dataset1 without outliers for price prediction with three different kernels.

The results are significantly better when the 6800 outliers were removed. RBF kernel works better this time. ME decreased to almost one third of its original

Table 5.2 Table of results for the regression model on price. Normalized error is calculated using the standard deviation method

Kernel	ME	MAE	MSE	RMSE	NMAE	NRMSE
Linear	−9.56	20.53	2082.15	45.63	41.8%	93.0%
RBF	−9.88	20.52	2133.56	46.19	41.8%	94.2%

Table 5.3 Table of results for the regression model on price, without outliers. Normalized error is calculated using the standard deviation method

Kernel	ME	MAE	MSE	RMSE	NMAE	NRMSE
Linear	−3.80	13.07	325.25	18.03	64.3%	88.7%
RBF	−3.75	12.94	318.64	17.85	63.6%	87.8%

evaluation. MAE and RMSE, which represents the same unit to the prediction values, are decreased close to 40% and 60%, respectively. The dramatic decrease of MSE indicates the outliers do not affect the results as before.

However, the NMAE increased, while, interestingly, the NRMSE decreased. This can be explained by the standard deviation being cut by nearly half, from around $40 with outliers to around $20 without outliers. The NMAE was not as sensitive to the removal of outliers as NRMSE, but both were equally sensitive to the change in standard deviation, leading to one increasing while the other decreased. Making the transition from a dataset with outliers to a dataset without outliers justifies the use of standard deviation for computing normalized error; the normalized minimax ($0–98) errors on the dataset without outliers would be around 13–17%. The large drop in price range drove up the normalized error, even though the model performs better without outliers.

In summary, predicting a wine's price through its review may not be so easy. Based on the MAR and RMSE value, there's a $13–18 differences between the real price versus the predicted price. One maybe curious why predicting the wine grade is better than predicting the wine price? This is because wine grade and wine reviews are tightly correlated since both came from wine reviewers; while wine prices reflect wine growers' and makers' hard work and wine reviewer reflects only the final product, there is a gap between attributes and target labels. In the end, wine reviewers decide the wine grade yet wine makers decide the wine price. Thus, this research work might be more likely to predict the wine reviewers' "expected price". Nonetheless, an error of $13–18 in the space of $0–98 still promising, considering this research might be the first research works on predicting a wine's price from its review in large scale of dataset.

References

1. Palmer, J., Chen, B.: Wineinformatics: regression on the grade and price of wines through their sensory attributes. Fermentation. **4**(4), 84 (2018)
2. Fradkin, D., Muchnik, I.: Support vector machines for classification. DIMACS Ser. Discret. Math. Theor. Comput. Sci. **70**, 13–20 (2006)
3. Martin, L.: A Simple Introduction to Support Vector Machines. Michigan State University, East Lansing, MI (2011)
4. Smits, G.F., Jordaan, E.M.: Improved SVM regression using mixtures of kernels. In Proceedings of the 2002 International Joint Conference on Neural Networks, Honolulu, HI, USA, 12–17 May 2002, vol. 3, pp. 2785–2790. IEEE, Washington, DC (2002)
5. R Core Team: R: A Language and Environment for Statistical Computing. R Foundation for Statistical Computing, Vienna (2015) https://www.R-project.org. Accessed 28 Sep 2018
6. Microsoft and R. C. Team: Microsoft R Open. Microsoft, Redmond, WA (2017) https://mran. microsoft.com/. Accessed 28 Sep 2018
7. Karatzoglou, A., Smola, A., Hornik, K., Zeileis, A.: Kernlab-an S4 package for kernel methods in R. J. Stat. Softw. **11**, 1–20 (2004)

Chapter 6
Multi-Class, Multi-Label and Multi-Target in Wineinformatics

Abstract *"Can wine grade, price and region being predicted altogether with higher accuracy?"* In previous chapters, bi-class classification and regression were discussed in grade class prediction, evaluation of wine reviewers and price prediction independently; that is, each classifier was given one task at a time. This chapter introduces an advanced computer science topic, multi-label and multi-target techniques, to Wineinformatics. Independent single-label problems, including wine grade, price and origin predictions, are merged together so that the trustworthy labels can also become attributes to provide more information for other labels.

6.1 Multi-Class, Multi-Label and Multi-Target

In general, a classification problem focuses on a single variable as the target for prediction. Different target value determines the type of classification; while The target value is a binary variable, a continuous value or one of multiple categories, the classification problem become bi-class classification, regression or multi-class classification, respectively. In other words, a **multi-class classification** tries to predict one target value as one of category with three or more choices. For example, the grade of a wine can be easily categorized into four classes: 100–95, 94–90, 89–85 and 84–80.

In this chapter, the experiments not only focuses on multi-class classification, also try to predict multiple response values; If **all** response values are in binary format, this is known as a **multi-label classification**. Multi-label classification can be formally defined as the problem of developing a model that maps input attributes x to **binary** vectors y as the labels for prediction. One of the most frequently used example for multi-label classification is movie genre prediction: one movie may have action, Sci-fi; the other movie may have action, drama and romance in their genre. Each genre is a label to the problem; so if a database holds ten different genres, there are ten labels to predict. Each genre is a binary label, if positive, the movie holds the genre; if negative, the movie does not have the genre.

If the prediction escalates to predict multiple response values and the response values are not all binary, which means they can be any combination of regression, binary classification, or multi-class classification, this problem is known as **multi-target prediction** [1]. If all the targets are categorical or discrete values, the problem become multi-target classification or multi-class multi-label classification; if all the targets are continuous values, the problem become multi-target regression.

6.2 Wine Labels for Multi-Target Prediction

The data source used in this work is still dataset1, The Big Dataset. Since the goal of this chapter is to predict wine region, grade and price simultaneously, the target labels need to be defined and transformed before the classification training process.

First, consider the **wine region** of origin. There are two wildly recognized styles of producing wine: an old world style and a new world style [2]. Old world wines mainly come from Israel and Europe, such as France, Spain, Italy. New world wines come from anywhere else, such as United States, Australia, Chile. The old and new world style tend to reflect the wine making techniques and where the wine was made. Therefore, if a wine is made in the old world, a value of "1" is assigned to region label; if a wine is made in the new world, a value of "0" is assigned. The design makes the region a bi-class label.

Next, the **wine grade** followed Wine Spectator's classification scheme and categorize the wine into "good", "very good", "outstanding", and "classic". This design makes wine grade a four classes label. The wine grade label can also easily transform back to a bi-class label by grouping "good, very good" and "outstanding, classic" as 89− and 90+, which was used in Chaps. 3 and 4.

Finally, the **wine price** utilized the boxplot analysis in the previous chapter to define categories. Quartiles were chosen to balance the wines evenly between wine groups: <$18, $18–29, $29–50, >$50; all prices were normalized to USD per 750 mL. This design makes wine price a four classes label. The wine price label can also easily transform back to a bi-class label by grouping "less than or equal to $29" and "greater than $29" as 29− and 29+. Tables 6.1 and 6.2 show the breakdown of the class labels for both multi-target and multi-label experiments,

Table 6.1 Class labels for multi-target experiment

Grade category	Grade	Price	Region
0	80–84	<$18	New world
1	85–89	$18–29	Old world
2	90–94	$29–50	
3	95–100	>$50	

Table 6.2 Class labels for multi-label experiment, where all labels have only binary classes

Grade category	Grade	Price	Region
0	80–89	≤$29	New world
1	90–100	>$29	Old world

Table 6.3 Percent of wines under $50 and over $200 in each wine grade category

Grade category	Percent of wine under $50	Percent of wine over $200
Good (80–84 pts)	95.89%	0.02%
Very Good (85–89 pts)	88.09%	0.14%
Outstanding (90–94 pts)	54.53%	2.82%
Classic (>95 pts)	15.51%	23.39%

where all labels are bi-class. That is to say, in a multi-target problem, if the wine has the label of [3, 1, 0], the wine has the grade greater than 95 pts, the price is between $18–29 and it from new world. In a multi-label problem, if the wine has the label of [1, 0, 1], which can also be presented as (grade, old world) with two positive attributes, the wine has the grade greater than 90 pts, the price is less than $29 and it from old world.

While analyzing wine prices, an interesting positive correlation between the grade classes and the price. Table 6.3 gives the clear picture to the correlation. In other words, if a wine is predicted as a class wine, the price of the wine might be on higher side, and vice versa. This observation provides an example of why predicting different labels together instead indecently might increase some or even all labels performances; since the earlier predicted label can serve as an attribute to other unpredicted labels, especially when the earlier predicted label has high performances.

6.3   How to Predict Multi-Label and Multi-Target Problems?

There are two general procedures to make multi-label and multi-target predictions: problem transformation and algorithm adoption. The first procedure applies transformation to the dataset so that the multi-label and multi-target problem becomes one or more single-label problems [3]. The second procedure indicates the classification algorithm is designed for multi-label and multi-target predictions directly. Nonetheless, the second procedure, algorithm adoption, actually use problem transformation within them [4]. Therefore, the core problem on multi-label and multi-target predictions still lies in how data is transformed.

6.3.1   Binary Relevance

Although there are three labels to be predicted simultaneously, each label can build their own model independently for prediction and serve as a baseline for comparison purposes; This is known as the **binary relevance** method. Consider dataset1, The Big Dataset, which contains 985 columns corresponding to normalized attributes

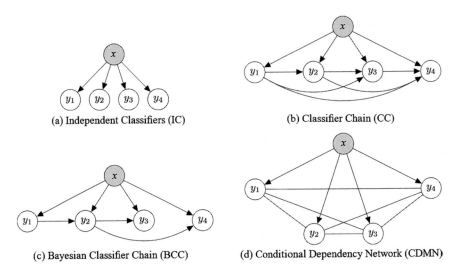

Fig. 6.1 (**a–d**) A graphical comparison of different multi-label methods [6]. Where x are the attributes, and y are the labels

plus additional three columns representing the labels in terms of grade, price **and** region. Assume all labels are binary, which means the problem is multi-label classification, the binary relevance method will create three datasets, each dataset will contain 985 columns corresponding to normalized attributes plus only one column representing the labels in terms of grade, price **or** region. Each dataset corresponds to the grade, price and region will be trained and evaluated independently. This approach is very efficient but might ignore correlations between labels [5]. If not all labels are binary, which means the problem is multi-target classification, rather than training a binary classifier on each label, a multi-class classifier is trained on the multi-class labels. Figure 6.1a represents the idea of binary relevance.

6.3.2   Label Powerset

The second problem transformation approach is the Label Powerset [3], which creates a new label that covers all possible combinations of the targeted labels. The total number of classes is equal to the number of classes from each dependent variable, multiplied together.

In the multi-label problem for grade, price and region, each label has two classes, so there would be a total of eight ($2 \times 2 \times 2$) classes for the new label. In the multi-target problem, since both grade and price have 4 classes and region has 2 classes, the label powerset method creates 32 ($4 \times 4 \times 2$) classes for the new label. Therefore, multi-label and multi-target problems are "transformed" into multi-class problem. A multi-class classifier is then trained on this new label for classification.

6.3.3   Classifier Chains

The third problem transformation approach is the classifier chains, which attempts to maintain the label dependency information while still using binary relevance method [7]. Classifier chains method has n stages, where n equals to the number of targeted labels. In the first stage of the classifier chains, a regular classifier is built for one label, just like a single label problem. The first classifier will be able to make predictions on the designated label. In the second stage of the classifier chains, the first stage predictions are used as an additional input to predict the second dependent variable [7]. In this way, the second stage classifier trains both all attributes and the class label predicted by the first classifier. By the same token, the third stage classifier trains with all attributes and the class label predicated by the first and the second classifiers, and so on. Figure 6.1b represents the idea of classifier chains.

To implement the classifier chain, except the first stage, there is a small difference between the training and testing process. In training, the second stage classifier uses the correct values of the first label; in testing, the second stage classifier takes the testing output from the first stage classifier to fill in the values of the first label. Since training process always take known and correct labels, the order of the label selection for each stage does not matter; however, it becomes obvious that the order in testing process does matter. Therefore, the first classifier should choose the label with the highest performance so that the predicted results have higher chance providing correct information for the following classifiers.

6.3.4   Bayesian Classifier Chains and Conditional Dependency Networks

The fourth problem transformation approach is the Bayesian classifier chains [8], which is similar to classifier chains but do not use all labels in the previous stages to classify the next staged label. The idea embedded in this method lies in not all labels are correlated to each other. The connection between classifiers can be determined by domain experts or some ensemble methods. The sparsity in the connections between labels leads to a faster classifier, and potentially a more accurate one, as labels which do not have dependencies between them need not be linked [6]. Figure 6.1c represents the idea of Bayesian classifier chains.

The last problem transformation approach is the conditional dependency networks, which is made by forming an undirected, fully connected graph between labels, as shown in Fig. 6.1d. Conditional dependency networks, unlike classifier chains or Bayesian classifier chains, does not look for dependency of one label on another. Instead, it encodes the dependencies between different labels. This is done by training as many binary classifiers as there are labels, where each classifier defines a conditional probability distribution on one label given all the other labels and the input attributes; Thus, there is a conditional probability distribution built for each label [9].

6.4   Evaluation Metrics

Since multi-label and multi-target classification are very different from bi-class classification, several new evaluations are needed to evaluate the results. The goal in this chapter is to predict right labels, including wine grade, price and region, for the wine. The first evaluation metric is **hamming loss**, which is the analog to error in a single-label problem [3] that averages over mistakes on individual labels. It is defined as:

$$HammingLoss = \frac{1}{|D|} \sum_{i=1}^{|D|} \frac{|Y_i \Delta Z_i|}{|L|}$$

where D is the multi-label dataset for evaluation, Y_i is the true label set to the ith observation in D, Z_i is the predicted label set of the ith observation in D, and L is the set of possible labels [4]. Δ stands for the symmetric difference of two sets, which is the XOR operation in Boolean logic. The lower this difference, the better the performance.

Another loss function that simply checks for entire correctness is **zero-one loss**. It is a measure based on exact matches between the true and predicted label set on all labels. For each observation, if the predicted labels do not match the true labels entirely, the loss for the observation is one; if the predicted labels do match the true labels, the loss for the observation is zero. Mathematically zero-one loss is defined as:

$$ZeroOneLoss = \frac{1}{|D|} \sum_{i=1}^{|D|} \begin{cases} 1, Y_i \neq Z_i \\ 0, Y_i = Z_i \end{cases}$$

Accuracy, precision, recall also measures the performances in this chapter with minor modifications:

$$Accuracy = \frac{1}{|D|} \sum_{i=1}^{|D|} \frac{|Y_i \cap Z_i|}{|Y_i \cup Z_i|}$$

$$Precision = \frac{1}{|D|} \sum_{i=1}^{|D|} \frac{|Y_i \cap Z_i|}{|Z_i|}$$

$$Recall \quad = \frac{1}{|D|} \sum_{i=1}^{|D|} \frac{|Y_i \cap Z_i|}{|Y_i|}$$

While accuracy calculates the average of the number of times that the true and predicted labels intersect divided by the number of true and predicted labels. Precision calculates the average of the number of times that the true and predicted labels intersect divided by the size of predicted set. Recall calculates the average of the number of times that the true and predicted labels intersect divided by the size of the true label set. The numerator in all these metrics counts the number of correctly classified labels for an instance. The F score, uses the same definition introduced before, combines both precision and recall into a single metric by taking the harmonic mean of both, equals to $2 \times (recision^*recall)/(precision + recall)$. Therefore, in a multi-label problem, if a wine is predicted as [1, 0, 1] as Z_i and the true label is [1, 0, 0] as Y_i; since there are two positive values in predicted value, $|Z_i| = 2$; since there is only one positive value in true label, $|Y_i| = 1$; and $|Y_i \cap Z_i| = 1$.

6.5   Implementation of Multi-Label and Multi-Target in Wineinformatics

Based on the previous research performances, SVM with linear and RBF kernels was chosen as the classifier in this work (Quadratic and Cubic kernels were tested, but the results were not satisfactory). For the multi-label approach, the LIBSVM was implemented within MEKA (a Multi-Label Extension for WEKA) [10, 11]. In all cases, C-classification was used, the cost of the constraint violation parameter was set to one, and **fivefold** cross validations were employed.

6.5.1   Apply Multi-Label on the Big Dataset

Table 6.4 provides the full results on all multi-label methods with two different kernels. All labels are binary. Although multi-label algorithms predict all labels simultaneously, the researchers can still look into individual label and determine the correctness of the prediction. Grade, price and region showed in the table are the per-label bi-class accuracy results predicted by each method; Hamming loss, 0/1 loss, accuracy and F-score are the multi-label evaluations mentioned above.

Since **binary relevance** built individual classifiers for each label, the results are considered as the baseline for comparison. The accuracy for the grade prediction is very close to 85%. However, one may be wondering why does the accuracy reported in Chap. 3 for the grade prediction using the same dataset is 86.3%, which is slightly higher? This is due to two possible reasons, for one, the SVM implemented in the different package. One used SVM-light and the other one used LIBSVM [12] where

Table 6.4 Results for multi-label classifications

Method	Kernel	Grade	Price	Region	Hamming	0/1 loss	Accuracy	F-score
Binary relevance	Linear	0.847	0.749	0.843	0.187	0.451	0.676	0.787
	RBF	0.832	0.741	0.829	0.199	0.472	0.655	0.768
Label powerset	Linear	0.849	0.753	0.851	0.182	0.423	0.691	0.794
	RBF	0.824	0.736	0.824	0.205	0.458	0.654	0.760
Classifier chain G, P, R	Linear	0.847	0.733	0.843	0.197	0.452	0.665	0.779
	RBF	0.832	0.723	0.829	0.205	0.474	0.642	0.759
Classifier chain R, G, P	Linear	0.847	0.749	0.843	0.187	0.451	0.677	0.787
	RBF	0.832	0.742	0.829	0.199	0.471	0.656	0.769
Bayesian classifier Chain	Linear	0.847	0.732	0.844	0.192	0452	0.665	0.779
	RBF	0.832	0.722	0.829	0.206	0.474	0.641	0.758
Conditional dependency Network	Linear	0.837	0.725	0.844	0.198	0.458	0.653	0.766
	RBF	0.812	0.710	0.830	0.216	0.486	0.624	0.737

basic parameters setup might be different, especially the algorithm was designed for multi-label purposes. The other reason is because of the fivefold cross validation contains randomness in splitting the whole dataset into smaller groups, which will cause minor differences in every experiment. The price prediction reaches 74.9%, which means the classifier can predict three out of four wines correctly on whether the wine cost more than $29. The prediction is very different from the previous chapter, since Chap. 5 predicts the exact cost of a wine using regression. The wine region is also very well predicted, 84.3% of wines can be distinguished where they came from based on their wine reviews. This result suggests that the characteristics of the wine from old world and new world are distinct. Hamming loss lower than 20% also suggests that the classifier can predict most of the labels correctly; while 0/1 loss lower than 50% indicates the classifier can predict all three labels correctly for more than half of the wines. Overall speaking, the performance is satisfactory.

Of all multi-label methods, **label powerset** performs the best. It is outperformed binary relevance in all evaluation metrics using linear kernel. 0.2%, 0.4% and 0.8% improvements on grade, price and region per-label prediction; 0.5%, 2.8%, 1.5% and 0.7% improvements on Hamming loss, 0/1 loss, accuracy and F-score evaluations. Because label powerset transforms the problem from a multi-label problem into a multi-class problem, the classifier is able to use correlating information from all three dependent variables at the same time. For example, if a particular independent variable is well correlated with grade and would predict a high quality wine, the label powerset approach is able to use these moderate correlations between price and grade to nudge that wine prediction towards a higher price, even if the other information about that wine is only weakly correlated to price directly.

Two combination of **classifier chains** were tested: The first chain order is Region, Grade, Price (R, G, P). The other order we compared is Grade, Price, Region (G, P, R). The results are very similar to binary relevance with slighter lower in price prediction; however, no significant differences in different orders used in classifier chains. This is likely because there is a moderate relationship between grade and price. In addition, the data has a high dimension in number of attributes, so the influence from a single extra dimension added on at each stage of the chain is dwarfed by the influence of all the other dimensions. Also, it is interesting to note that the performance in price was less than binary relevance or label powerset, while the other two labels remained competitive.

The **Bayesian classifier chain** improved on the classifier chain was able to have almost identical performance with the baseline results. The order of the chain did not matter: The algorithm was able to identify the correct chain order regardless of our settings, and the performance was identical no matter the settings we used for chain order. Because of this, Bayesian classifier chain was able to outperform the standard classifier chain in the multi label metrics. **Conditional dependency networks** take a lot of time for training; however, the performance does not comparable with other methods. The grade and price predictions are 1% and 2.4% lower than binary relevance and worse in all multi-label evaluations.

6.5.2   Apply Multi-Target on the Big Dataset

Table 6.5 provides the full results on all multi-target methods with two different kernels. Both grade and price have four classes and region has two classes. Since accuracy and F-score are designed for multi-label problems, both evaluations are not applicable in this subsection.

Table 6.5 Results for multi-target classifications

Method	Kernel	Grade	Price	Region	Hamming	0/1 loss
Class relevance	Linear	0.752	0.470	0.844	0.312	0.697
	RBF	0.702	0.449	0.829	0.340	0.737
Label powerset	Linear	0.583	0.298	0.849	0.423	0.818
	RBF	0.544	0.247	0.821	0.463	0.864
Classifier chain	Linear	0.752	0.463	0.843	0.314	0.693
G, P, R	RBF	0.702	0.432	0.829	0.346	0.738
Classifier chain	Linear	0.752	0.464	0.844	0.314	0.692
R, G, P	RBF	0.702	0.435	0.829	0.344	0.736
Bayesian classifier	Linear	0.752	0.473	0.844	0.311	0.694
Chain	RBF	0.702	0.449	0.830	0.340	0.736
Conditional dependency	Linear	0.739	0.459	0.843	0.319	0.694
Network	RBF	0.680	0.420	0.830	0.357	0.744

Class relevance, which is the multi-target extension to binary relevance, still serves as the baseline results for discussion and comparison. Since region prediction is still a bi-class classification problem, the results stay the same with previous table. For the grade to be predicted in four classes, the accuracy is still as high as 75.1%; while the accuracy for the price prediction is 47.1%. For comparison to random chances, the expected accuracy in four classes is 25%. The hamming loss is about 31% and 0/1 loss is close to 70%.

Label powerset, unfortunately, does not perform as good as multi-target experiments. This is due to the increased number of classes that must be formed from the greater number of label combinations found in the four-class dataset. In total, $4*4*2 = 32$ class labels were generated. When there are more classes, the probability of error increases. While LP performed the best with two classes, it now performs the worst with multiple classes. Even so, it has the highest performance in predicting region.

Classifier chains continue to perform well when given the multi-target dataset. Like in the two-class problem, the order of the classifier chain shows some, but minor, differences, with a chain order of region, grade, and price beating the chain order of grade, price, and region in zero-one loss only by tenths of a percent.

The **Bayesian classifier chain** was able to outperform the regular classifier chain. Based on the results of our research, the order of the chain set by parameters has little influence in this dataset. Although the zero-one loss is slightly higher compared to regular classifier chains, the per-label accuracy and Hamming loss are improved.

The **conditional dependency network** has average results compared to the other methods. Although not as good as the classifier chain in per-label accuracies, the classifier trellis got very close in Hamming loss and zero-one loss, especially with the linear kernel. Despite the CDN having a poorer Hamming loss, it is able to remain competitive in zero-one loss.

In summary, this chapter demonstrates a state-of-the-art data science problem, multi-label and multi-target classification, applied on Wineinformatics research. While label powerset performs the best for the two-class dataset, the BCC remains the most reliable overall. The experimental results show that the multi-label multi-target algorithms has the potential to improve per-label accuracy by incorporating correlation between labels.

References

1. Spyromitros-Xioufis, E., Groves, W., Tsoumakas, G., Vlahavas, I.: Multi-label classification methods for multi-target regression. arXiv, 1211.6581v1 (2012)
2. Anderson, K.: The World's Wine Markets: Globalization at Work. Edward Elgar, Cheltenham (2004)
3. Tawiah, C.A., Sheng, V.S.: Empirical comparison of multi-label classification algorithms. In: Proc. 27th AAAI Conf. on Artificial Intelligence, Bellevue, WA, USA, pp. 2–6. AAAI, Menlo Park, CA (2013)

4. Tsoumakas, G., Katakis, I.: Multi-label classification: an overview. In: Erickson, J. (ed.) Database Technologies: Concepts, Methodologies, Tools, and Applications, pp. 4–6. IGI Global, Barcelona (2009) 10–12

5. Zhang, M.-L., Li, Y.-K., Liu, X.-Y., Geng, X.: Binary relevance for multi-label learning: an overview. Front. Comput. Sci. **12**(2), 191–202 (2018)

6. Read, J., Martino, L., Olmos, P.M., Luengo, D.: Scalablemulti-output label prediction: from classifier chains to classifier trellises. Pattern Recogn. **48**(6), 2096–2109 (2015)

7. Read, J., Pfahringer, B., Holmes, G., Frank, E.: Classifier chains for multi-label classification. Mach. Learn. **85**(3), 333–359 (2011)

8. Zaragoza, J.H., Sucar, L.E., Morales, E.F., Bielza, C., Larrañaga, P.: Bayesian chain classifiers for multidimensional classification. In: Proc. 22nd Int. Joint Conf. on Artificial Intelligence, Barcelona, Spain, pp. 2192–2197. AAAI, Menlo Park, CA (2011)

9. Guo, Y.H., Gu, S.C.: Multi-label classification using conditional dependency networks. In: Proc. 22nd Int. Joint Conf. on Artificial Intelligence, Barcelona, Spain, pp. 1300–1305. AAAI, Menlo Park, CA (2011)

10. Hall, M., Frank, E., Holmes, G., Pfahringer, B., Reutemann, P., Witten, I.H.: The WEKA data mining software: an update. ACM SIGKDD Explor. Newslett. **11**(1), 10–18 (2009)

11. Read, J., Reutemann, P., Pfahringer, B., Holmes, G.: MEKA: a multi-label/multi-target extension to Weka. J. Mach. Learn. Res. **17**(1), 667–671 (2016)

12. Chang, C.C., Lin, C.J.: LIBSVM: a library for support vector machines. ACM Trans. Intell. Syst. Technol. **2**(3), 27 (2011)

Chapter 7
Advanced Usage of the Computational Wine Wheel

Abstract *"How can computers understand wine reviews even more?"* Since the Computational Wine Wheel has built-in hierarchy among attributes; this chapter wants to address how to extract even more information from wine reviews through the Computational Wine Wheel to represent CATEGORY and SUBCATEGORY information. Unlike Chaps. 4–6 that use Dataset 1 for the experiments, two wine region specific datasets from Bordeaux in France are used to show how to retrieve regional information from targeted area.

7.1 Additional Information Captured by the Computational Wine Wheel

The Computational Wine Wheel introduced in Chap. 2 contains 14 attributes in the CATEGORY, 34 in the SUBCATEGORY, 1881 in the ORIGINAL words, and 985 in the NORMALIZED attributes. Throughout all other chapters, the wine reviews were mapped with the ORIGINAL words and captured NORMALIZED attributes as binary information for the datasets creation. However, the CATEGORY and SUBCATEGORY information were not utilized at all. For example, for a wine with "fresh apple" and "crushed cherry" in the review, the computational wine wheel will encode a 1 for the "apple" attribute since apple is the normalized attribute for "fresh apple" and a 1 for the "cherry" attribute since cherry is the normalized attribute for "crushed cherry." In this chapter, not only "apple" and "cherry" attributes will be 1, the new data will count "tree fruit" twice and "fruity" twice since both apple and cherry are under the "tree fruit" SUBCATEGORY and "fruity" CATEGORY columns. Therefore, the new dataset changes from a purely binary dataset to a mixed data format dataset, which will increase the complexity for the computational costs and hopefully also increase the quality of the retrieved information and performance. The flowchart presented in Fig. 7.1 shows the process of converting reviews into a machine readable format. Reviews are processed using the computational wine wheel into NORMALIZED, SUBCATEGORY, and CATEGORY_NAME.

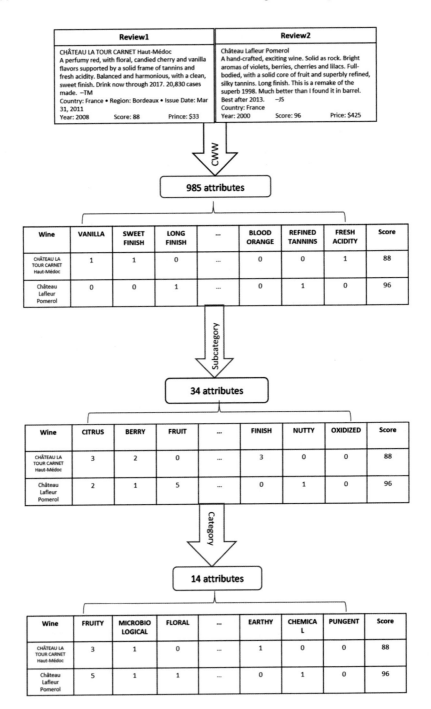

Fig. 7.1 The process of using the Computational Wine Wheel to retrieve NORMALIZED, SUBCATEGORY, and CATEGORY_NAME attributes [1]

Table 7.1   attributes selected for each sub-dataset of dataset 2 and 3

Dataset2	Dataset3	
	Elite	
Bordeaux	Bordeaux	D
2.1	3.1	14 category attributes
2.2	3.2	34 subcategory attributes
2.3	3.3	985 normalized attributes
2.4	3.4	14 category attributes + 34 subcategory attributes + 985 normalized attributes
2.5	3.5	14 category attributes + 34 subcategory attributes
2.6	3.6	34 subcategory attributes + 985 normalized attributes
2.7	3.7	14 category attributes + 985 normalized attributes

14 CATEGORY and 34 SUBCATEGORY attributes were available by counting the occurrence of the mapped NORMALIZED attributes. However, both CATEGORY and SUBCATEGORY attributes are continuous instead of binary values with varied range; without normalize the data, the classification algorithm may give more weights to those newly added attributes. Therefore, Min-Max normalization were used to scale the attribute values from 0 to 1. The Min-Max normalization formula is:

$$x' = \frac{x - \min(x)}{\max(x) - \min(x)}$$

where x is the original value, x' is the normalized value, min(x) and max(x) are the minimum and maximum value among all x. Once all attributes are generated and normalized, it is time to answer the questions: "Will the newly generated category and subcategory attribute help in building more accurate classification models?" and "How can a system be designed to integrate three different types of attributes?"

Since there are three set of attributes, seven sub-datasets were built to test all possible combinations based on Dataset2 twenty-first Century Bordeaux dataset with 14,349 wines and Dataset3 twenty-first Century Elite Bordeaux dataset with 1359 wines. These sub-datasets are defined as follows: sub-datasets 1, 2, and 3 reflect the three individual sets of attributes. Sub-dataset 4 encompasses all three sets of attributes and sub-datasets 5, 6, and 7 are constructed of a combination of two out of the three individual sets. Details of each dataset can be found in Table 7.1.

Please note that the dataset 2.3 and 3.3 are the very datasets used in Chap. 3. Therefore, results generated from dataset 3 is the comparison results with previous research work. Dataset 2.4 and 3.4 are the complete dataset using all attributes.

7.2   Naïve Bayes Classification Algorithm for Mixed Datatype

As mentioned in Chap. 3, Naive Bayes classifier is considered as the most suitable white-box classification algorithm in Wineinformatics. The Bayes' theorem can be defined as:

$$P(c|x) = \frac{P(x|c)P(c)}{P(x)}$$

Where x be a testing wine sample that the classifier wants to make the prediction; Let c be a hypothesis (our prediction) that x belongs to class C; Classification is to determine P(c|x), the probability that the hypothesis holds given the observed data sample x; P(c) (prior probability), the initial probability for the class; P(x): probability that sample data is observed; P(x|c) (posteriori probability), the probability of observing the sample x, given that the hypothesis holds. $P(x|c) = P(x1|c) * P(x2|c)*$ ... $P(xn|c)$ where $x1, x2, \ldots xn$ represent each attribute of x. In other words, $P(x1|c)$ means what is the probability of attribute $x1$ occurs in each class in the training dataset.

In Chap. 3, **Bernoulli** Naïve Bayes classifier with Laplace smoothing, which is designed for binary attributes, were introduced with a concrete example; In this chapter, since attributes from CATEGORY and SUBCATEGORY are continuous values, **Gaussian** Naïve Bayes classifier were implemented for these attributes [1, 2]. In Gaussian Naive Bayes classifier, it is assumed that the continuous values associated with each attribute are distributed according to a Gaussian distribution. A Gaussian distribution is also known as Normal distribution.

$$P(X_i|Y) = \frac{1}{\sigma_y\sqrt{2\pi}} \exp\left(-\frac{(X_i - \mu_y)^2}{2\sigma_y^2}\right)$$

where μ_y is the sample mean, σ_y is the sample standard deviation. When a value of X never appears in the training set, the prior probability of that value of X will be 0. If we do not use any techniques, $P(Y| X_1, X_2, \ldots, X_n)$ will be 0, even when some of the other prior probability of X are very high. This case does not seem fair to other X. Therefore, the smallest μ_y and σ_y among all attributes to the X were assigned to handle zero prior probability. Naïve Bayes classifier makes the assumption of there is no dependence between attributes and the importance of each attribute is the same. Therefore, for the dataset with mixed data types, including dataset 2.4–2.7 and 3.4–3.7, the probability for continuous values and binary values were calculated independently and multiply together for the final class probability.

SVM was designed for both binary and continuous variables, therefore, there is no modification necessary for the new datasets. All experiments results reported in the next section were evaluated through evaluation metrics discussed in Chap. 3 with fivefold cross validation.

7.3   Does Additional Attributes Help?

7.3.1   Case Study on the Twenty-First Century Bordeaux Dataset

Naïve Bayes classifier was applied on the dataset 2.1–2.7 with encouraging results showed on Table 7.2.

According to the table, datasets 2.1, 2.2 and 2.5 did not perform as good as other datasets since the number of attributes included in the dataset are much lower than others. Compare with the baseline dataset 2.3, although using all attributes can be extracted from the Computational Wine Wheel, dataset 2.4 did not perform as well in terms of accuracy, recall and F-score. This observation suggests that more attributes do not necessary provide better information for the Naïve Bayes classification model. However, dataset 2.7, which combines 14 Category and 985 normalized attributes, generated the best results in three out of four evaluation metrics. These results suggest that attributes from Subcategory do not increase the information quality in the dataset2 using the Naïve Bayes algorithm.

To compare with Naïve Bayes classifier, Table 7.3 shows the results generated from SVM on datasets 2.1–2.7.

Table 7.2 Accuracy, precision, recall, and F-score with Naïve Bayes classifier in Dataset 2, highest results in each column are marked in bold

Dataset 2	Accuracy	Precision	Recall	F-score
2.1	74.39%	61.64%	36.48%	45.83%
2.2	74.72%	61.17%	40.86%	48.98%
2.3	85.17%	73.22%	**79.03%**	76.01%
2.4	82.37%	77.65%	57.10%	65.80%
2.5	74.93%	62.09%	40.11%	48.73%
2.6	84.79%	81.32%	63.38%	71.22%
2.7	**87.32%**	**81.94%**	73.52%	**77.49%**

Table 7.3 Accuracy, precision, recall, and F-score with SVM classifier in Dataset 2, highest results in each column are marked in bold

Dataset 2	Accuracy	Precision	Recall	F-score
2.1	80.46%	73.31%	53.81%	62.06%
2.2	82.09%	75.58%	58.67%	66.06%
2.3	86.97%	**80.68%**	73.80%	77.10%
2.4	**87.00%**	80.35%	74.50%	**77.31%**
2.5	82.12%	75.53%	58.88%	66.17%
2.6	**87.00%**	80.31%	**74.53%**	77.30%
2.7	86.92%	80.13%	74.45%	77.18%

In this table, all accuracies are higher than 80%, even for the datasets with less attributes. These results are very exciting, especially dataset 2.1 contains only 14 attributes. Compared with the dataset 2.3, the datasets 2.4 and 2.6 performed slightly better in accuracy, recall as well as F-score. However, the improvement is not obvious. These results suggest that attributes from Category and Subcategory do slightly increase the information quality in the All Bordeaux datasets using SVM.

Comparing the results given in both tables, the best performance came from the dataset 2.7 using Naïve Bayes classification algorithm. This result is encouraging since SVM, a black box classification algorithm, usually performs better than the Naïve Bayes algorithm. This means Naïve Bayes can capture extra information from 14 category attributes to build a more accurate, even better than SVM, classification model.

7.3.2   Case Study on the Twenty-First Century Elite Bordeaux Dataset

Naïve Bayes classifier was also applied on the dataset 3.1–3.7 with encouraging results showed on Table 7.4.

Except for datasets 3.1, 3.2 and 3.5, all other datasets with additional information performed better than the previous results from dataset 3.3, which severed as the baseline performances. The dataset 3.4 achieved the highest accuracy, recall and F-score; the results suggest that additional information from CATEGORY and SUBCATEGORY does help the Naïve Bayes algorithm to build the classification model. The 94.33% recall indicates that more than 94% of the wines in the 1855 Bordeaux category can be recognized by our classification research. In terms of precision, the dataset 3.7 achieved as high as 91.34%, which has never been achieved before.

Table 7.5 shows the results generated from SVM on datasets 3.1–3.7.

The dataset 3.4 achieved the highest results in all evaluation metrics. The consistent results produced from both tables suggests that the smaller dataset, the 1855 Bordeaux dataset, benefits more from the full power of the Computational Wine Wheel. These results suggest that the dataset 3.4 is the best processed data for understanding 90+ points twenty-first century elite Bordeaux wines. Both Naïve

Table 7.4 Accuracy, precision, recall, and F-score with Naïve Bayes classifier in Dataset 3, highest results in each column are marked in bold

Dataset 3	Accuracy	Precision	Recall	F-score
3.1	70.36%	77.15%	78.21%	77.37%
3.2	71.97%	75.10%	85.73%	79.91%
3.3	84.62%	86.79%	90.02%	88.38%
3.4	**86.18%**	85.92%	**94.33%**	**89.89%**
3.5	76.02%	76.08%	92.63%	83.43%
3.6	86.17%	86.54%	93.31%	89.78%
3.7	85.88%	**91.34%**	86.72%	88.84%

Table 7.5 Accuracy, precision, recall, and F-score with SVM classifier in Dataset 3, highest results in each column are marked in bold

Dataset 3	Accuracy	Precision	Recall	F-score
3.1	71.22%	73.94%	86.17%	79.55%
3.2	79.18%	82.17%	86.74%	84.37%
3.3	81.38%	86.84%	84.12%	85.46%
3.4	**86.38%**	**89.42%**	89.68%	**89.53%**
3.5	85.58%	86.47%	**92.29%**	89.26%
3.6	82.48%	86.67%	86.28%	86.46%
3.7	85.57%	89.05%	88.77%	88.89%

Bayes and SVM produced outstanding results. Unlike results shown in Table 7.3, SVM does catch more information from additional attributes resulting in an increase of 5.00% and 4.19% accuracy in the 3.4 and 3.7, respectively. Interestingly, the dataset 3.5, which uses only CATEGORY and SUBCATEGORY (total of 48 attributes), performed better than the baseline dataset 3.3 with the highest recall. This finding indicates that the SVM put more emphasis on the CATEGORY and SUBCATEGORY attributes.

In deciding which classification algorithm performed better, F-score will be the tie breaker to indicate Naïve Bayes slightly outperformed SVM. On top of that, Naïve Bayes is a white-box classification algorithm, so the results can be interpreted. Therefore, based on the findings in this paper, Naïve Bayes is the best classification algorithm for Wineinformatics research, utilizing the full power of the Computational Wine Wheel.

7.3.3   The More Attributes, the Better?

Dataset 2.4 and 3.4 contains 1033 attributes (14 category attributes + 34 subcategory attributes + 985 normalized attributes) and dataset 2.7 and 3.7 contains 999 attributes (14 category attributes + 985 normalized attributes). Experimental results suggest that the dataset using 1033 and 999 attributes outperformed the dataset using 985 attributes, which were used in previous Wineinformatics researches. However, drawing the conclusion "the more attributes, the better" is still not very clear. To visualize the experimental results using datasets 2.4, 3.4 and 2.7 and 3.7; Fig. 7.2 is created based on Tables 7.2 and 7.3; Fig. 7.3 is created based on Tables 7.4 and 7.5.

In Fig. 7.2, all results are very similar except dataset 2.4 using Naïve Bayes. More specifically, the recall performed poorly. Dataset2 contains a total of 14,349 wines with 4263 90+ wines and 10,086 89− wines. This imbalanced situation might be the cause for Naïve Bayes to perform poorly in dataset 2.4, which means the 34 attributes from SUBCATEGORY cannot provide positive impacts to the model generated from All Bordeaux wines.

In Fig. 7.3, in terms of accuracy and F-score, both the datasets 3.4 and 3.7 performed very similar under the SVM and Naïve Bayes algorithms. The only result that needs to be addressed is the recall from the dataset 3.44 using Naïve Bayes,

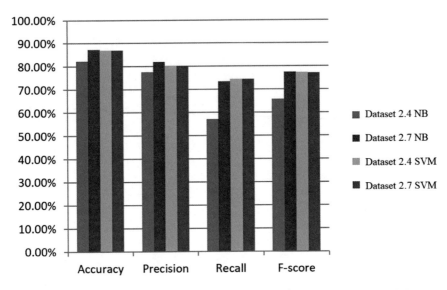

Fig. 7.2 Graphical comparison of datasets 2.4 and 2.7 using Naïve Bayes (NB) and SVM

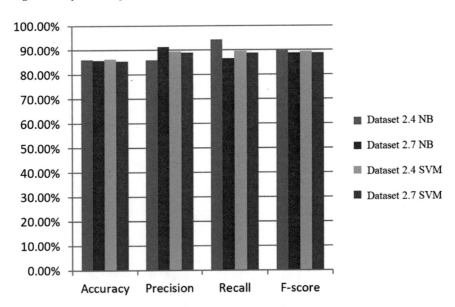

Fig. 7.3 Graphical comparison of datasets 3.4 and 3.7 using Naïve Bayes (NB) and SVM

which reaches as high as 94%. This result might be affected by the same imbalanced situation of the dataset: in the 1855 Bordeaux dataset with 1359 collected wines, there are 882 90+ wines and 477 89− wines. Unlike the data distribution of the first dataset, which has much more 89− wines than 90+ wines, the dataset3 has more 90+ wines than 89− wines. In either case, this research suggests that attributes retrieved from CATEGORY and SUBCATEGORY have the power to provide more information to classifiers for superior model generation. This finding provides strong impact to many Wineinformatics research efforts in different wine related topics, such as evaluation on wine reviewers [3], wine grade and price regression analysis [4], terroir study in a single wine region [5], weather impacts examination [6], and multi-target classification on a wine's score, price and region [7].

References

1. Dong, Z., Atkison, T., Chen, B.: Wineinformatics: using the full power of the computational wine wheel to understand 21st century bordeaux wines from the reviews. Beverages. **7**(1), 3 (2021)
2. Lou, W., Wang, X., Chen, F., Chen, Y., Jiang, B., Zhang, H.: Sequence based prediction of DNA-binding proteins based on hybrid feature selection using random forest and Gaussian naive Bayes. PLoS One. **9**, e86703 (2014)
3. Chen, B., Velchev, V., Palmer, J., Atkison, T.: Wineinformatics: a quantitative analysis of wine reviewers. Fermentation. **4**, 82 (2018)
4. Palmer, J., Chen, B.: Wineinformatics: regression on the grade and price of wines through their sensory attributes. Fermentation. **4**, 84 (2018)
5. Chen, B., Velchev, V., Nicholson, B., Garrison, J., Iwamura, M., Battisto, R.: Wineinformatics: uncork napa's cabernet sauvignon by association rule based classification. In: Proceedings of the 2015 IEEE 14th International Conference on Machine Learning and Applications (ICMLA), Miami, FL, USA, 9–11 December 2015, pp. 565–569. IEEE, Washington, DC (2015)
6. Chen, B., Jones, D., Tunc, M., Chipolla, K., Beltrán, J.: Weather impacts on wine, a BiMax examination of napa cabernet in 2011 and 2012 vintage. In: Proceedings of the ICDM 2019, New York, NY, USA, 17–21 July 2019, pp. 242–250. IEEE, Washington, DC (2019)
7. Palmer, J., Sheng, V.S., Atkison, T., Chen, B.: Classification on grade, price, and region with multi-label and multi-target methods in wineinformatics. Big Data Min. Anal. **3**, 1–12 (2019)

Chapter 8
Conclusion and Future Works

8.1 Conclusion

In this book, **Wineinformatics**, a new data science application domain, was introduced systematically for the first time. Human language format wine reviews served as the main data source for this work. In order to accurately pre-process wine review data, the **Computational Wine Wheel** was utilized as the human language processing tool to extract not only flavors but also non-flavor attributes. Three dataset with different purposes were carefully developed and analyzed in different chapters to answer important questions in Wineinformatics. Chapter 1 gave a general introduction to the book. Chapter 2 discussed about the data source, the Computational Wine Wheel and datasets creation. Chapter 3 introduced classification, algorithms and evaluations to the topic.

Starting from Chap. 4, novel research ideas in Wineinformatics were presented in each chapter. In Chap. 4, nine reviewers from Wine Spectator were evaluated based on the consistency of grading wines 90+ or 89− based on more than 100,000 wines. Reviewers were ranked according to the performances of different classification models. In Chap. 5, regression models were tested on predicting actual wine grade and price from wine reviews. The results suggested that wine grades can be successfully predicted within 1.6 points errors; while wine prices had a $13–18 differences between the real price versus the predicted price. The gap indicates there is a difference between wine reviewers' expected wine prices versus wine makers' suggested market prices.

Chapter 6 brought the research complexity in Wineinformatics into next level; Multi-class, Multi-label and Multi-target were introduced to this new data science research domain. Wine grade, price and region can be predicted and evaluated using different Multi-label and Multi-target algorithms. Subtle correlation between different labels were captured through classification models. The experimental results show that the multi-label multi-target algorithms has the potential to improve per-label accuracy by incorporating correlation between labels. Chapter 7 tried to include new CATEGORY and SUBCATEGORY information from the process of using the Computational Wine Wheel. While the data type became heterogeneous data type that mixed binary and continuous values, higher performances for classification models were observed. These results provide new perspective in data preprocessing and strong impact to many Wineinformatics research efforts in different wine related topics in included in this book and many other Wineinformatics topics and future works.

Due to space limitation, this book was focused on classification problems in Wineinformatics. All other learning type of algorithms including unsupervised learning, semi-supervised learning, and reinforced learning can be easily adopted into this novel data science field. Unsupervised learning, or clustering, can group similar wines together and provide recommendations to users for food paring or purchasing strategies [1–3]. For example, a relatively lower cost wine among a group of high end and high cost wines can be suggested as an entry level wine for consumers. Semi-supervised learning merges the power of supervised learning and unsupervised learning for more precise classification results [4]. Reinforced learning, such as neural networks and deep learning, can be applied to Wineinformatics for machines to understand and evaluate wines.

8.2   Future Works

Many new directions and novel researches can be studied in the study of Wineinformatics. First of all, including new wine reviews from different wine experts may create the next version of the Computational Wine Wheel to increase the variety and complexity of natural language processing works. Different wine reviewers use different style of reviewing writing; Robert Parker, one of the most famous wine reviewers in the world, has his reviews described as "remarkably powerful contemporary rhetoric which has had an unprecedented impact in the world of prestigious wine for more than two decades" [5]; therefore, unlike Wine Spectator's reviews, which are shorter and more precise, Robert parker's reviews are more descriptive and colorful, and they provide greater challenges in natural language processing.

Many other beverages use similar review systems, such as beer [6], whiskey [7], tea and coffee [8] and share similar keywords in their reviews. However, different beverages include diverse topics that may be convertible through the basic version of the Computational Wine Wheel. For example, beer reviews use 1/3 of paragraph to discuss about the foam of the beer. Additional human language processing works need to be done for creating beverage-specific modules to be included in the Computational Wine Wheel, which can be evolved into the Computational FLAVOR wheel.

Last but not least, the ultimate goal of Wineinformatics is to use AI and computers to objectively review wines with the help of chemical compound analysis. Wine reviewers may not always agree with each other since they have their own personal preferences as shown in Chap. 4. Even if they agree on the overall grade of the wine, the words they describe the same wine are usually very different. Chateau Latour 2009 was one of the most "perfect" wines in wine history, all Robert Parker, James Suckling, Wine Enthusiast and Wine Spectator gave it 100 or 99 points as the final verdict; however, the reviews they provided are very different from each other [9]. Therefore, one of the major research direction in Wineinformatics is to combine wine reviews from different sources and merge wine reviewers' knowledge in wine to create an objective wine reviewer system. The chemical compound analysis can be utilized as the wine expert tasting process as the source of the wine data and map it to the objective wine reviewer system for grading and knowledge extraction.

References

1. McCune, J., Riley, A., Chen, B.: Clustering in wineinformatics with attribute selection to increase uniqueness of clusters. Fermentation. **7**(1), 27 (2021)
2. Chen, B., Rhodes, C., Alexander, Y., Velchev, V.: The computational wine wheel 2.0 and the TriMax triclustering in wineinformatics. In: Perner, P. (ed.) Industrial Conference on Data Mining, pp. 223–238. Springer, Cham (2016)
3. Chen, B., Rhodes, C., Crawford, A., Hambuchen, L.: Wineinformatics: applying data mining on wine sensory reviews processed by the computational wine wheel. In: 2014 IEEE International Conference on Data Mining Workshop, pp. 142–149. IEEE, Washington, DC (2014)
4. Chen, B., Buck, K.H., Lawrence, C., Moore, C., Yeatts, J., Atkison, T.: Granular computing in wineinformatics. In: 2017 13th International Conference on Natural Computation, Fuzzy Systems and Knowledge Discovery (ICNC-FSKD), pp. 1228–1232. IEEE, Washington, DC (2017)
5. Hommerberg, C.: Persuasiveness in the Discourse of Wine: The Rhetoric of Robert Parker. Ph.D. Thesis, Linnaeus University Press, Kalmar, Sweden (2011)
6. Craft beer and brewing. https://beerandbrewing.com/. Accessed 21 Jun 2022
7. Whiskey Advocate. https://www.whiskyadvocate.com/ratings-and-reviews/. Accessed 21 Jun 2022
8. Coffee Review. https://www.coffeereview.com/. Accessed 21 Jun 2022
9. Wine. https://www.wine.com/product/chateau-latour-15-liter-magnum-2009/725922. Accessed 21 Jun 2022

Printed in the United States
by Baker & Taylor Publisher Services